Wissenschaftliche Re
Fahrzeugtechnik Uni

Reihe herausgegeben von
Michael Bargende, Stuttgart, Deutschland
Hans-Christian Reuss, Stuttgart, Deutschland
Jochen Wiedemann, Stuttgart, Deutschland

Das Institut für Verbrennungsmotoren und Kraftfahrwesen (IVK) an der Universität Stuttgart erforscht, entwickelt, appliziert und erprobt, in enger Zusammenarbeit mit der Industrie, Elemente bzw. Technologien aus dem Bereich moderner Fahrzeugkonzepte. Das Institut gliedert sich in die drei Bereiche Kraftfahrwesen, Fahrzeugantriebe und Kraftfahrzeug-Mechatronik. Aufgabe dieser Bereiche ist die Ausarbeitung des Themengebietes im Prüfstandsbetrieb, in Theorie und Simulation. Schwerpunkte des Kraftfahrwesens sind hierbei die Aerodynamik, Akustik (NVH), Fahrdynamik und Fahrermodellierung, Leichtbau, Sicherheit, Kraftübertragung sowie Energie und Thermomanagement – auch in Verbindung mit hybriden und batterieelektrischen Fahrzeugkonzepten. Der Bereich Fahrzeugantriebe widmet sich den Themen Brennverfahrensentwicklung einschließlich Regelungs- und Steuerungskonzeptionen bei zugleich minimierten Emissionen, komplexe Abgasnachbehandlung, Aufladesysteme und -strategien, Hybridsysteme und Betriebsstrategien sowie mechanisch-akustischen Fragestellungen. Themen der Kraftfahrzeug-Mechatronik sind die Antriebsstrangregelung/Hybride, Elektromobilität, Bordnetz und Energiemanagement, Funktions- und Softwareentwicklung sowie Test und Diagnose. Die Erfüllung dieser Aufgaben wird prüfstandsseitig neben vielem anderen unterstützt durch 19 Motorenprüfstände, zwei Rollenprüfstände, einen 1:1-Fahrsimulator, einen Antriebsstrangprüfstand, einen Thermowindkanal sowie einen 1:1-Aeroakustikwindkanal. Die wissenschaftliche Reihe „Fahrzeugtechnik Universität Stuttgart" präsentiert über die am Institut entstandenen Promotionen die hervorragenden Arbeitsergebnisse der Forschungstätigkeiten am IVK.

Reihe herausgegeben von

Prof. Dr.-Ing. Michael Bargende
Lehrstuhl Fahrzeugantriebe
Institut für Verbrennungsmotoren und Kraftfahrwesen, Universität Stuttgart
Stuttgart, Deutschland

Prof. Dr.-Ing. Hans-Christian Reuss
Lehrstuhl Kraftfahrzeugmechatronik
Institut für Verbrennungsmotoren und Kraftfahrwesen, Universität Stuttgart
Stuttgart, Deutschland

Prof. Dr.-Ing. Jochen Wiedemann
Lehrstuhl Kraftfahrwesen
Institut für Verbrennungsmotoren und Kraftfahrwesen, Universität Stuttgart
Stuttgart, Deutschland

Weitere Bände in der Reihe http://www.springer.com/series/13535

Bülent Sari

Fail-operational Safety Architecture for ADAS/AD Systems and a Model-driven Approach for Dependent Failure Analysis

Springer Vieweg

Bülent Sari
Institute of Internal Combustion Engines
and Automotive Engineering (IVK)
University of Stuttgart
Stuttgart, Germany

Zugl.: Dissertation Universität Stuttgart, 2019

D93

ISSN 2567-0042 ISSN 2567-0352 (electronic)
Wissenschaftliche Reihe Fahrzeugtechnik Universität Stuttgart
ISBN 978-3-658-29421-2 ISBN 978-3-658-29422-9 (eBook)
https://doi.org/10.1007/978-3-658-29422-9

This Springer Vieweg imprint is published by the registered company Springer Fachmedien Wiesbaden GmbH part of Springer Nature.
The registered company address is: Abraham-Lincoln-Str. 46, 65189 Wiesbaden, Germany

Preface

This work was realized during my tenure as a research associate at the Institute of Internal Combustion Engines and Automotive Engineering (IVK) at the University of Stuttgart under the supervision of Prof. Dr.-Ing. H.-C. Reuss.

My deep gratitude goes to Prof. Dr.-Ing. H.-C. Reuss for his valuable hints and his trust in my abilities. I also thank Prof. Dr.-Ing. Eric Sax from Karlsruhe Institute of Technology (KIT) for the second referring.

I am extremely grateful to all of my colleagues at ZF Friedrichshafen AG for their support and encouragement, especially my supervisors Roland Geiger and Frank König.

This thesis has been supported by the Master thesis of Fabio Blaschke and the Internship of Preetham Kamsagara Madegowda. I would like to thank them and also proofreaders Alexander Achleitner, Roland Geiger and Laura Ebersberger for improving this thesis.

My appreciation also extends to my company ZF Friedrichshafen AG for the financial support to attend the national and international conferences while writing my thesis.

Finally, I would like to thank my wife and children for their support and understanding during this time.

Friedrichshafen Buelent Sari

Contents

Figures

Tables

Abbreviations

ADAS	Advanced Driver Assistance Systems
AD	Autonomous Driving
ADC	Analog to Digital Converter
AEB	Automated Emergency Braking
ASIL	Automotive Safety Integrity Level
ATESST	Advancing Traffic Efficiency and Safety through Software Technology
CCU	Capture Compare Unit
CPU	Central Processing Unit
CRC	Cyclic Redundancy Check
DFA	Dependent Failure Analysis
DFI	Dependent Failure Initiators
DGPS	Differential Global Positioning System
DMA	Direct Memory Access
DOORS	Dynamic Object Oriented Requirements System
DSL	Domain-specific Language
EAST-ADL	Electronics Architecture and Software Technology - Architecture Description Language
EEA	Embedded Electronic Architecture
EDC	Error Detection Code
ECC	Error Correction Code
ECU	Electronic Control Unit
E/E/PES	Electrical/Electronic/Programmable Electronic Safety-related Systems
EUC	Equipment Under Control
FAA	Functional Analysis Architecture
FDA	Functional Design Architecture
FFI	Freedom from Interference
FMEA	Failure Modes and Effects Analysis
FOV	Field of view
FTA	Fault Tree Analysis

FKFS	Forschungsinstitut für Kraftfahrwesen und Fahrzeugmotoren Stuttgart
HARA	Hazard analysis and risk assessment
HDA	Hardware Design Architecture
HiP-HOPS	Hierarchically Performed Hazard Origin and Propagation Studies
GPT	General Purpose Timer
GPU	Graphics Processing Unit
GTM	Generic Timer Module
GUI	Graphical User Interface
IMU	Inertial Measurement Unit
IVK	Institut für Verbrennungsmotoren und Kraftfahrwesen
LiDAR	Light detection and ranging
MAENAD	Model-based Analysis and Engineering of Novel Architectures for Dependable Electric Vehicles
MPU	Memory Protection Unit
NHTSA	National Highway Traffic Safety Administration
OICA	Organisation Internationale des Constructeurs d'Automobiles
OEMs	Original Equipment Manufacturers
Radar	Radio Detection and Ranging or Radio Direction and Ranging
ReqIF	Requirements Interchange Format
SAE	Society of Automotive Engineers
SAFE	Safe Automotive Software Architecture
SDA	Software Design Architecture
SMU	Safety Management Unit
SoC	System-on-a-Chip
SOTIF	Safety of the Intended Functionality
SRAM	Static Random-access Memory
SysML	Systems Modeling Language
TIM	Timers
TMR	Triple Modular Redundancy
1oo2	One-out-of-Two
2oo2D	Two-out-of-Two-Diagnosis
2oo3	Two-out-of-Three

Abstract

The self-driving automotive technology has been progressing very rapidly in the last few years. Many automotive companies that are developing autonomous vehicles want to launch them with SAE level 4 at the beginning of 2020. The main goal in the development of self-driving cars is to reduce accidents caused by driver errors. But there are still some technological challenges to be solved, such as the safety increase and driving availability necessary to gain customer acceptance. The primary purpose of this research is to investigate the possible fail-operational safety architectures for conventional systems with a powertrain, and for the automated driving system in consideration of the entire ADAS/AD processing chain, from complex sensors to perception and decision algorithms. The solutions developed in this work show how a redundant system architecture and safety architecture can be created efficiently, and how diverse redundancy can be designed in ADAS/AD systems including the processing chain to fulfill the ASIL-D safety requirements and increase the system availability with fail-operational design for self-driving vehicles with SAE Level 3 and fully autonomous vehicles with SAE level 4 and level 5.

The increased functionality of vehicle systems, resulting from the electrification of the power train and required by autonomous driving, leads to greater complexity in the design of the system, hardware, software and safety architecture. At the same time, the use of multicore processors in the automotive industry is becoming necessary because of the need for more processing power, more memory and higher safety requirements. The safety goals for most automotive items are classified with ASIL D, especially for automated driving. Therefore, it is necessary to investigate the safety solutions particularly for systems rated with the ASIL-D. The ISO 26262 provides the possibility to apply a decomposition approach for safety requirements. An appropriate decomposition has the advantage of reducing the ASIL rating of the top-level safety requirements, but the use of ASIL decomposition requires redundancy of safety requirements. These redundant safety requirements shall be allocated to sufficiently independent architectural elements. A dependent failure analysis (DFA) provides evidence that the decomposed function parts are adequately independent of one another. The secondary purpose of the research is to develop an approach for model-based dependent failure analysis which is required by ISO 26262 for safety systems which use the decomposition approach. This extended model-based DFA method also considers SOTIF-related technological shortcomings such as sensor limitations due to weather conditions. While common cause failures and cascading failures are already investigated for independence, the SOTIF-related triggering events should also be evaluated regarding the independence of the decomposition paths.

Additionally, the model-driven approach for the development of the main activities from ISO 26262, such as hazard analysis and risk assessment, functional safety concept, technical safety concept, safety analysis, etc., is discussed briefly. The extensions and developed scripts make it possible to gain sufficient transparency and traceability for the safety arguments and to support the entire safety process in a single solution throughout the hardware and software architectural development.

Kurzfassung

Die selbstfahrende Fahrzeugtechnologie hat in den letzten Jahren sehr schnell Fortschritte gemacht. Viele OEMs und Zulieferer investieren für deren autonome Systeme und wollen damit die autonomen Fahrzeuge am Anfang 2020 mit SAE-Level 4 auf den Markt bringen. Bei der Entwicklung selbstfahrender Autos geht es vor allem darum, Unfälle durch Fahrfehler zu reduzieren. Dafür gibt es aber noch einige technologische Herausforderungen wie Erhöhung der Sicherheit und Verfügbarkeit, die für Erhöhung der Kundenakzeptanz mit innovativen Lösungen gemeistert werden sollen. Das Hauptziel dieser Forschungsarbeit ist die Untersuchung der möglichen fehlertoleranten Sicherheitsarchitekturen für konventionelle Systeme mit einem Antriebsstrang und für die automatisierten Fahrzeugsysteme unter Berücksichtigung der gesamten ADAS/AD Verarbeitungskette, von komplexen Sensoren wie Kamera, Radar, Lidar, etc., bis hin zu Wahrnehmungs- und Entscheidungsalgorithmen. Die Lösungen zeigen, wie die redundante Systemarchitektur und Sicherheitsarchitektur effizient erstellt werden kann und wie diversitäre Redundanz in ADAS/AD-Systemen einschließlich der Verarbeitungskette zur Erfüllung der ASIL-D Sicherheitsanforderungen gestaltet werden kann und wie die Systemverfügbarkeit mit fehlertoleranten Design für selbstfahrende Fahrzeuge mit SAE Level 3 und voll autonome Fahrzeuge mit SAE Level 4 und Level 5 erhöht werden kann.

Die erhöhte Funktionalität von Fahrzeugsystemen, die aus der Elektrifizierung des Antriebsstrangs resultiert und durch autonomes Fahren erforderlich ist, führt zu einer größeren Komplexität bei der Erstellung der System-, Hardware-, Software- und Sicherheitsarchitektur. Gleichzeitig wird der Einsatz von Multicoreprozessoren in der Automobilindustrie aufgrund des Bedarfs an mehr Rechenleistung, mehr Speicher und höheren Sicherheitsanforderungen notwendig. Die Sicherheitsziele sind aktuell für die meisten Automobilsysteme mit ASIL D klassifiziert, insbesondere für das automatisierte Fahren. Daher ist es notwendig, die Sicherheitslösungen insbesondere für Systeme zu untersuchen, die für das ASIL-D ausgelegt sind. Die ISO 26262 bietet die Möglichkeit, einen Dekompositionsansatz für höhere Sicherheitsanforderungen anzuwenden. Eine entsprechende Dekomposition hat den Vorteil, die höchsten Sicherheitsanforderungen auf mehrere redundante Sicherheitsanforderungen zu zerlegen und deren ASIL-Klassifizierung zu reduzieren. Aber die Voraussetzung dafür ist, dass diese redundanten Sicherheitsanforderungen zu den ausreichend unabhängigen architektonischen Elementen zugeordnet werden müssen. Eine Unabhängigkeitsanalyse (DFA) liefert den Nachweis, dass die zerlegten Funktionsteile hinreichend unabhängig voneinander sind. Das zweite Ziel der Forschung besteht darin, einen Ansatz für automatisierte modellbasierte DFA zu entwickeln, deren Durchführung gemäß ISO 26262 für Sicherheitssysteme bei der Anwendung von Dekompositionsansatz erforderlich ist. Diese erweiterte modellbasierte DFA-Methode berücksichtigt auch SOTIF-bezogene technologische Unzulänglichkeiten, wie z.B. Sensoreinschränkungen aufgrund von Wetterbedingungen. Während Gemeinsame-Fehler (Common-cause) und Kaskaden-Fehler bereits auf ihre Unabhängigkeit hin untersucht werden, sollen die SOTIF-bezogenen Triggerereignisse auch hinsichtlich der Unabhängigkeit der Dekompositionspfade ausgewertet werden.

Darüber hinaus wird der modellgetriebene Ansatz zur Entwicklung der Hauptaktivitäten aus ISO 26262 wie Gefahren- und Risikoanalyse, funktionales-/technisches Sicherheitskonzept, Sicherheitsanalyse, etc., kurz diskutiert. Die Erweiterungen und entwickelten Automatisierungsskripte ermöglichen eine ausreichende Transparenz und Nachvollziehbarkeit für die Sicherheitsargumente und unterstützen den gesamten Sicherheitsprozess in einer einzigen Lösung auch in der System-, Hardware- und Software-Architekturentwicklung.

1 Introduction

1.1 Motivation and Objectives

According to the global status report on road safety [12], 1.2 million people die each year due to traffic accidents around the world. Driver error is the cause of 94 % of the accidents which occur in the USA [11]. Of these driver-related critical errors, 41 % are recognition errors, 33 % are decision errors, 11 % are performance errors and 7 % are non-performance errors (sleep, etc.). Reducing the number of accidents is a great challenge and also an ethical mission for the automotive industry. This challenge can only be mastered with the development of Advanced Driver Assistance Systems (ADAS) and Autonomous Driving (AD). These aspects must be considered in the design of vehicle systems to improve safety. Therefore, many companies are investing in the development of their ADAS and fully self-driving systems [23, 24, 48].

It has also become necessary in recent years, for the automotive industry to use multicore processors and domain ECUs. These technologies bring definite advantages such as greater processing power, more memory and added safety features, and the use of these technologies has increased with the ingress of E-mobility and autonomous driving.

The number of ADAS/AD systems in road vehicles is increasing. These systems are classified into six levels of driving automation, from "no automation" to "full automation", which are identified and described in the new SAE International standard J3016 [13]. SAE Level 1 means that only one driver assistance system for either steering or acceleration/deceleration can be executed, while SAE Level 2 means that more driver assistance systems for both longitudinal and lateral control can be executed. SAE Level 3 vehicles can execute all driving functions and can have control under certain circumstances with the expectation that if the system requests it, the human driver will take over the control. SAE Level 4 vehicles are able to drive autonomously in some driving modes considering all aspects of the dynamic driving task without expecting

B. Sari, *Fail-operational Safety Architecture for ADAS/AD Systems and a Model-driven Approach for Dependent Failure Analysis*, Wissenschaftliche Reihe Fahrzeugtechnik Universität Stuttgart, https://doi.org/10.1007/978-3-658-29422-9_1

any driver intervention. SAE Level 5 vehicles are able to drive autonomously in all situations and driving environments, and, additionally, SAE Level 4 and 5 systems should also be able to bring the vehicle to a safe state or to remain fail-operational in the case of any system failure without the intervention of a human driver.

According to ISO 26262 [14], the safety violations that result from malfunctions of the E/E system and random hardware failures in vehicles, and also according to SOTIF [15], safety violations in an E/E fault free system that result from faulty sensor data and also processing algorithms due to nominal performance of sensors and algorithms, shall be avoided or mitigated to bring the system to a safe state. This means that in the case of a fault, the system must either be switched off or be driven in a degradation mode. For SAE Level 3 and especially for SAE Level 4, innovative solutions are necessary for functional safety, system availability and system redundancy regarding fail-operational. According to SAE Level 3, the driver cannot instantaneously take over control of the vehicle and also, (according to SAE Level 4) the driver cannot be considered as a system fallback. Therefore, the system must ensure safety for the period of time that the driver is still not engaged, and, in case of SAE Level 4 and Level 5, while the driver is not available. This makes the fail-operational systems essential for autonomous driving.

The primary purpose of this research is to discuss the possibilities for fail-operational safety architecture for ADAS systems and to examine AD systems considering the processing chain from sensors to the domain ECUs with high performance chips.

In a premium vehicle, up to 100 ECUs are installed, which are capable of computing complex algorithms. Overall, the embedded software of a premium automobile contains up to 100 million lines of source code. As a comparison, the new "Boeing Dreamliner 787" needs only about 6.5 million lines of code for all onboard systems [20]. This comparison shows the complexity of the systems in modern vehicles. This complexity will continue growing. The reason for this extensive requirement for automotive systems is the electrification of the automobile and ADAS/AD systems. It is believed that the ratio of electronic components to the total production cost of a vehicle could rise to up to 35 % by 2020 and up to 50 % by 2030 [1]. With the increase of electrification, the proportion of safety-critical systems also increases. The malfunctions; excess deceleration, inadvertent lateral control, and unintended deployment of the airbag while driving, are a few examples that can lead to life-threatening injuries and are therefore classified with ASIL-D.

The ISO 26262 provides the possibility to apply the decomposition approach to safety critical systems. An appropriate decomposition has the advantage of decreasing the ASIL rating of the top level safety requirement derived from safety goal into the redundant safety requirements and thus enables a reduction of the development effort. But the use of ASIL decomposition requires the independence of redundant safety requirements, and therefore these safety requirements must be allocated to sufficiently independent architectural elements. To apply the decomposition strategy, ISO 26262 requires that dependent failure analysis (DFA) is performed, thus providing evidence for sufficient independence between decomposed function parts. Currently, it is possible to use commercial external tools [5] to perform the safety analysis within EAST-ADL. These tools are capable of automatically generating a fault tree analysis (FTA) and a failure modes and effects analysis (FMEA) from an EAST-ADL error model. But at this time, DFA is performed manually and there is no supporting tool for that. This results in additional development effort, because the whole path from system decomposition down to software and hardware decompositions must be analyzed to ensure that the signals and hardware parts are sufficiently independent. The other disadvantage of manual analysis is that it is difficult to achieve traceability.

The secondary purpose of this research is to develop an approach for the model-based DFA that is required by ISO 26262 for safety systems with the decomposition approach. This extended model-based DFA method also considers SOTIF-related technological shortcomings such as sensor limitations due to weather conditions.

In addition, this research discusses how model-driven approach can be applied to the safety development cycle from ISO 26262. The modeling of a complete system is an extensive project that requires domain knowledge. Various tools know-how is also needed, because the architecture is described using several tools. An alternative to the current approach is a domain-specific language based on modified EAST-ADL. With this extended DSL, it is possible that all information is created within a system model that describes the complete system, safety, hardware and software architecture.

The extensions and developed scripts make it possible to gain sufficient transparency and traceability for the safety arguments in consideration of multicore processors and enable support of the entire safety process in a single solution, throughout the system, safety, hardware and software architecture development.

1.2 Thesis Outline

This thesis consists of five chapters and is structured as shown in the following figure:

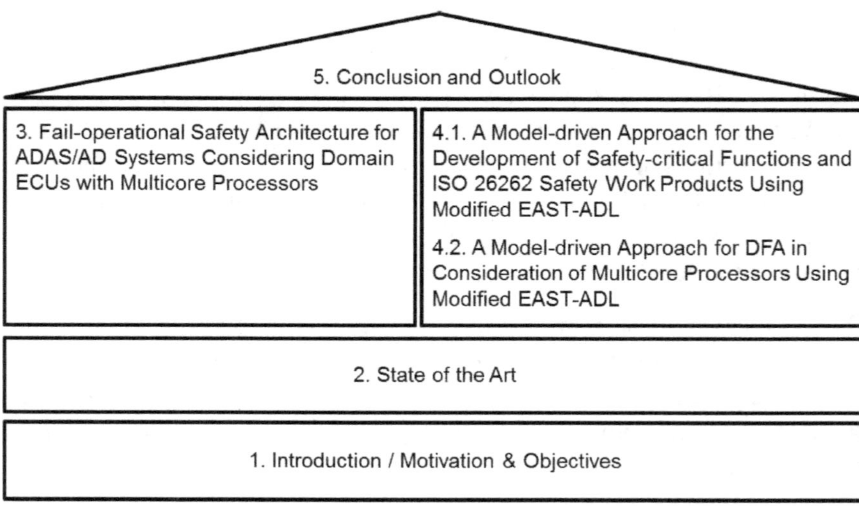

Figure 1.1: Structure of the thesis

Chapter 2 covers the related state-of-the-art on functional safety, and the relevant aspects of ISO 26262 such as the safety lifecycle, ASIL decomposition, DFA and FFI. In addition, ISO/PAS 21448 (SOTIF), SAE J3016 (Automated

Driving Levels), multicore processors, domain ECUS and EAST-ADL are briefly described.

Chapter 3 describes the fail-operational safety architectures for vehicle systems considering multicore processors and domain ECUs which contain separate multicore processors. Fail-operational safety architectures are investigated for both ADAS/AD systems and conventional systems such as powertrain. The focus of section 3.4 is on the fail-operational safety architectures for ADAS/AD systems considering processing chain and domain ECUs with multicore processors. Besides the development of various fail-operational safety architectures, this subchapter investigates the solution using ASIL decomposition for ADAS/AD systems and also the use of the DFA considering also SOTIF-related triggering events. The related research results are published in [37, 38, 41, 44].

Chapter 4 presents two model-driven approaches. The first one concerns the model-based development of safety-critical functions and ISO 26262 work products using modified EAST-ADL standard. The second approach is about the model-driven approach for DFA in consideration of multicore processors. The research results related to this topic are published in [36, 39, 40, 42, 43].

Finally, Chapter 5 concludes the results and contributions of this thesis.

2 State of the Art

2.1 Functional Safety

Safety is becoming more important with the increasing level of safety-related E/E Systems built into cars. The increasing functionality of vehicle systems through the electrification of the powertrain and autonomous driving technology leads to more complexity in the design of the system, hardware, software and safety architecture. To handle this increased complexity and develop the safer systems systematically, functional safety standards are being developed.

The following Fig. 2.1 shows the history of the development of functional safety standards:

	IEC 61508 2004	**ISO 26262 – First Edition** 2011	**ISO 26262 – Second Edition** 2018
Functional Safety Standard	**Functional Safety of Electrical/Electronic/Programmable Electronic Safety-related Systems (E/E/PE, or E/E/PES).**	**Road Vehicles - Functional safety – First edition**	**Road Vehicles - Functional safety – Second edition**
Scope	**applicable to all kinds of industry**	**applicable to all road vehicles less than 3.5 tons**	**applicable to all road vehicles / commercial vehicles / motorcycles**

Figure 2.1: Safety standard history

IEC 61508 is an international functional safety standard published by the International Electrotechnical Commission and it is titled "Functional Safety of

B. Sari, *Fail-operational Safety Architecture for ADAS/AD Systems and a Model-driven Approach for Dependent Failure Analysis*, Wissenschaftliche Reihe Fahrzeugtechnik Universität Stuttgart, https://doi.org/10.1007/978-3-658-29422-9_2

Electrical/Electronic/Programmable Electronic Safety-related Systems (E/E/PE or E/E/PES)" [6].

The IEC 61508 was developed as a basic functional safety standard applicable to all types of industry. It defines functional safety as: "part of the overall safety relating to the EUC (Equipment Under Control) and the EUC control system which depends on the correct functioning of the E/E/PE safety-related systems, other technology safety-related systems and external risk reduction facilities"[6].

The ISO 26262 was developed on the basis of IEC 61508 to handle the specific needs of electrical and/or electronic (E/E) systems within road vehicles of less than 3.5 tons, excluded mopeds and special vehicles such as E/E systems designed for drivers with disabilities.

The scope of the ISO 26262 has been extended to all road vehicles including commercial vehicles and motorcycles in the second edition of the standard.

According to ISO 26262, functional safety is defined as follows [14]:

"absence of unreasonable risk due to hazards caused by malfunctioning behavior of E/E systems",

where unreasonable risk is defined as "risk judged to be unacceptable in a certain context according to valid societal moral concepts" and malfunctioning behavior is defined as "failure or unintended behavior of an item with respect to its design intent" and risk is defined as "combination of the probability of occurrence of harm and the severity of that harm" [14].

The purpose of functional safety is to avoid the unreasonable risks caused by systematic failures and random hardware failures of E/E systems.

2.2 ISO 26262 - Road Vehicles - Functional Safety

The safety-related systems, which are used in road vehicles and which consist of at least one electrical or electronic (E/E) system, should be developed according to the functional safety standard ISO 26262. ISO 26262 is a modification of IEC 61508 to address the safety-related automotive domains. This standard is applicable to all safety activities during the development lifecycle of safety-related items.

ISO 26262 addresses possible hazards caused by safety-related E/E hardware failure or unintended system behavior due to specification error or software failure. The hazards related to electric shock, fire, smoke, heat, radiation, toxicity, flammability, reactivity, corrosion, release of energy and similar hazards which are not caused directly by malfunctioning behavior of safety-related E/E systems, are not in the scope of ISO 26262.

The importance of functional safety is increasing according to the level of system complexity caused by the rising number of safety-related E/E Systems built into cars. This system complexity increases the probability of occurrence of harm from systematic failures and random hardware failures. ISO 26262 provides requirements to mitigate these risks in order to reduce them to an acceptable level.

In addition to the technical safety requirements to the system, hardware and software development, ISO 26262 provides guidance for safety management and development process requirements such as requirements for specification, design, implementation, integration, verification, validation and configuration, and the production and service processes.

The ISO 26262 describes the following tasks at the beginning of each part to achieve functional safety [14]. ISO 26262:

a "provides a reference for the automotive safety lifecycle and supports the tailoring of the activities to be performed during the lifecycle phases, i.e., development, production, operation, service, and decommissioning"

b "provides an automotive-specific risk-based approach to determine automotive safety integrity levels (ASIL)"

c "uses ASILs to specify which of the requirements of ISO 26262 are applicable to avoid unreasonable residual risk"

d "provides requirements for functional safety management, verification, validation and confirmation measures" e “provides requirements for relations with suppliers”

Fig. 2.2 shows the overall structure of ISO 26262 second edition. The second edition of ISO 26262 contains two additional parts. Part 11 provides a guideline on application of ISO 26262 to semiconductors and part 12 deals with the adaptation of ISO 26262 for motorcycles. ISO 26262 is based on the V-model as a reference process model for the different phases of product development [14]:

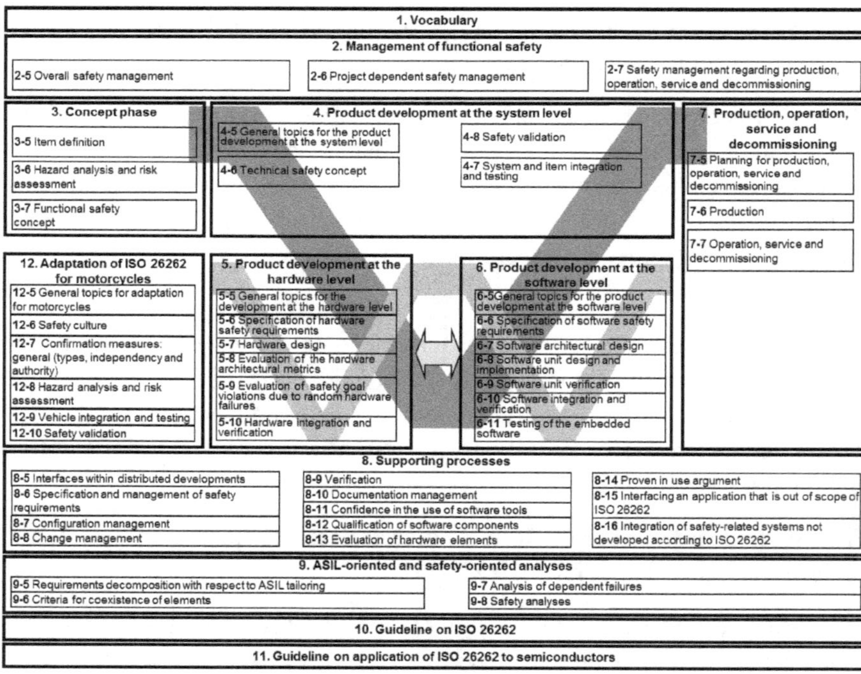

Figure 2.2: Overall structure of ISO 26262 [14]

2.2.1 Safety Lifecycle

Fig. 2.3 from ISO 26262-part2, clause5, shows the safety lifecycle which illustrates the key safety activities during the concept phase, product development, production, operation, service and decommissioning. These safety activities are documented within the safety work products which are planned in the safety plan at the beginning of the project.

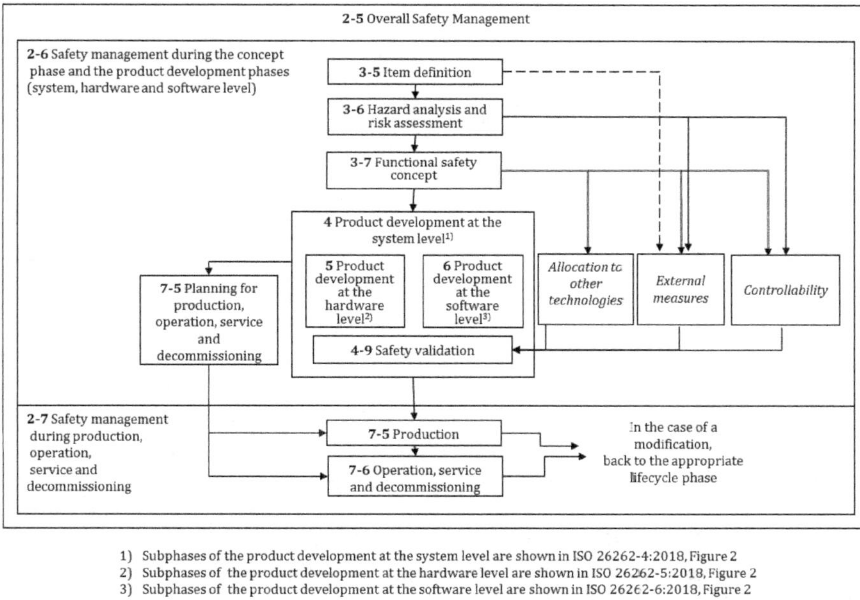

Figure 2.3: Safety lifecycle [14]

The first task, after the creation of the safety plan, is the item definition. The item definition encompasses the description of the item in consideration of its functionality, boundary, interfaces to other items, environmental conditions, etc. [14].

After the planning of the safety lifecycle and the definition of the item, the hazard analysis and risk assessment (HARA) is performed according to ISO 26262-part3, clause6. For this analysis, the relevant scenarios for the item are defined and their probability of exposure is estimated. The malfunctions of the item are then analyzed within these scenarios with consideration of the controllability and the severity of the related hazardous events. The ASIL rate of the hazardous events is classified within these parameters, namely Exposure (E), Severity (S) and Controllability (C). Finally, the safety goals for each ASIL classified hazardous event are determined. Table 2.1 shows the simple representation of the S, E, C classes. ISO 26262-part3, annexB gives a detail explanation about ASIL classification.

Table 2.1: Classes of exposure, severity and controllability [14]

Classes of probability of exposure				
E0	**E1**	**E2**	**E3**	**E4**
Incredible	Very low probability	Low probability	Medium probability	High probability
Classes of controllability				
	C0	**C1**	**C2**	**C3**
	Controllable in general	Simply controllable	Normally controllable	Difficult to control or uncontrollable
Classes of severity				
	S0	**S1**	**S2**	**S3**
	No injuries	Light and moderate injuries	Severe and life-threatening injuries (survival probable)	Life-threatening injuries (survival uncertain), fatal injuries

The safety goals are the top-level safety requirements of the item for mitigating the hazardous events to an acceptable level. The safety goals become the highest ASILs from the corresponding hazardous events. The classification of hazardous events is shown in the following table 2.2 from ISO 26262 [14].

A functional safety concept required by ISO 26262-part3, clause7 is developed in consideration of preliminary architectural assumptions based on the safety goals. The functional safety concept contains the safety goals as top level safety requirements and the corresponding functional safety requirements derived from safety goals. The functional safety requirements inherit the ASIL from the corresponding safety goal. After the functional safety concept is specified in the concept phase as given in ISO 26262-part3, the item is developed further at the system level, as given in ISO 26262-part4.

The system development process is based on the concept of a V-model and contains the safety work products technical safety concept, system architectural design, system design and implementation on the left side, and system and item integration and verification, safety validation and the functional safety assessment on the right side.

Fig. 2.4 from ISO 26262-part4 illustrates the subphases of the system development.

Table 2.2: ASIL determination [14]

Severity class	Exposure class	Controllability class		
		C1	C2	C3
S1	E1	QM	QM	QM
	E2	QM	QM	QM
	E3	QM	QM	A
	E4	QM	A	B
S2	E1	QM	QM	QM
	E2	QM	QM	A
	E3	QM	A	B
	E4	A	B	C
S3	E1	QM	QM	A
	E2	QM	A	B
	E3	A	B	C
	E4	B	C	D

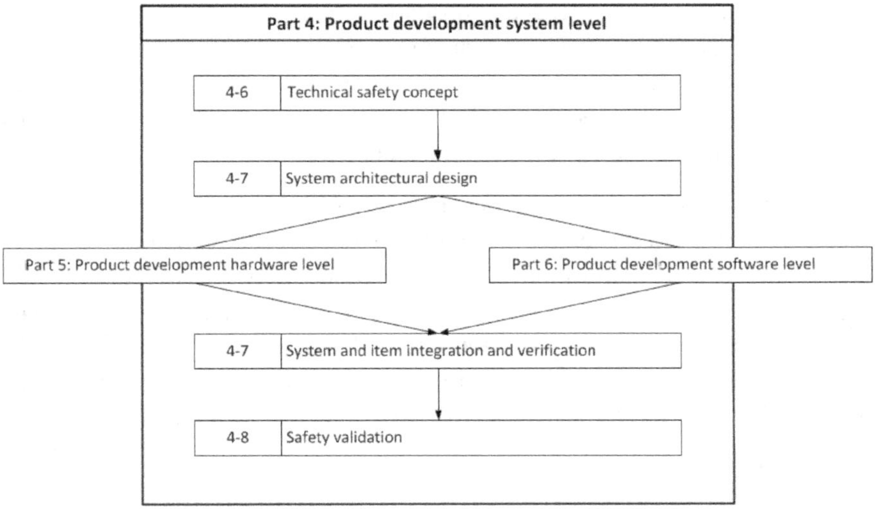

Figure 2.4: System development lifecycle [14]

The hardware is developed based on the system design specification and technical safety concept according to ISO 26262-part5. ISO 26262-part5, clause5.2 illustrates the reference phase model for the product development at the hardware level.

The software safety requirements are derived from functional safety concept and technical safety concept. To achieve the safety goals, the implementation of the safety functions is based on the system design specification according to ISO 26262-part6. ISO 26262-part6, clause5.2 shows the reference phase model for the software development.

2.2.2 ASIL Decomposition

In general, the safety goals should be achieved with the implementation of the corresponding safety requirements. These safety requirements should inherit the ASIL of the safety goals. In fact, the safety goals for the most important automotive items, e.g., powertrain, braking and steering are classified with the highest ASIL, namely ASIL D. It is usual to implement these safety goals with the decomposed safety requirements, because the ISO 26262 requirements such as diverse programming for ASIL D software demand high development costs and expenditure of time.

ISO 26262 defines the ASIL decomposition as follows:
"ASIL decomposition is the apportioning of redundant safety requirements to elements, with sufficient independence, conducing to the same safety goal, with the objective of reducing the ASIL of the redundant safety requirements that are allocated to the corresponding elements" [14].

As defined above, the usage of decomposition offers the opportunity to decrease the ASIL rating of the decomposed safety requirements of a safety goal. To apply decomposition, the decomposed safety requirements should be allocated to sufficiently independent architectural elements. If the redundant/decomposed safety requirements cannot be allocated to sufficiently independent architectural elements, then these redundant safety requirements inherit the initial ASIL of the safety goal. ASIL decomposition can be applied to the functional, technical, hardware or software safety requirements of the item.

The possible choice of decomposition schemes is outlined according to the ASIL of the safety goal, as follows [14]:

a An ASIL D requirement shall be decomposed as follows:

- ASIL C(D) requirement and ASIL A(D) requirement
- ASIL B(D) requirement and ASIL B(D) requirement
- ASIL D(D) requirement and ASIL QM(D) requirement

b An ASIL C requirement shall be decomposed as follows:

- ASIL B(C) requirement and ASIL A(C) requirement
- ASIL C(C) requirement and ASIL QM(C) requirement

c An ASIL B requirement shall be decomposed as follows:

- ASIL A(B) requirement and one ASIL A(B) requirement
- ASIL B(B) requirement and one QM(B) requirement

NOTE: An ASIL A requirement shall not be further decomposed. But if needed, then it can be decomposed as one ASIL A(A) requirement and one QM(A) requirement.

Further requirements for using decomposition are described in ISO 26262–part9, clause5.

The next chapter describes the analysis of dependent failures which is the most important aspect and requirement to prove the absence of dependent failures for the application of decomposition.

2.2.3 Analysis of Dependent Failures (DFA)

The purpose of dependent failure analysis is to find out single causes and failure modes that could violate a safety requirement by preventing the fulfillment of a required independence or a freedom from interference requirement between architectural elements. As given in ISO 26262-part9, clause7, independence of architectural elements is threatened by common cause failures and cascading failures, while freedom from interference is only violated by cascading failures as shown in Fig. 2.5. Dependent failure analysis should be performed when using the decomposition approach.

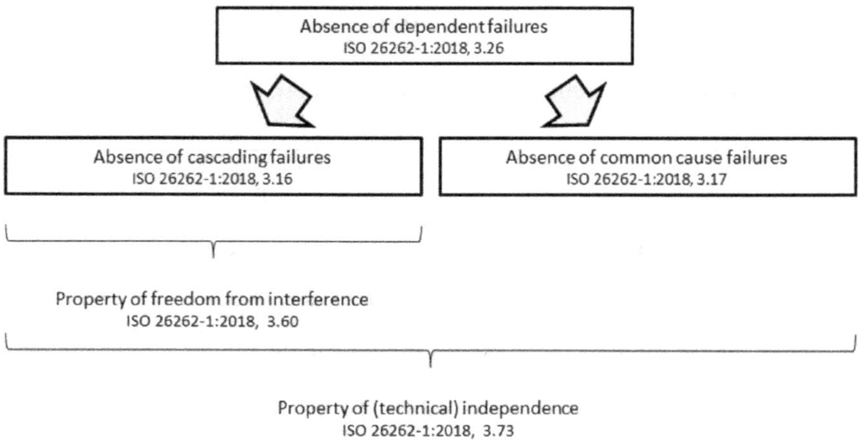

Figure 2.5: Dependent failure analysis (DFA) [14]

According to ISO 26262-part9, clause7, the analysis of dependent failures considers the following architectural features [14]:

- "similar and dissimilar redundant elements"
- "different functions implemented with identical software or hardware elements"
- "functions and their respective safety mechanisms"
- "partitions of functions or software elements"
- "physical distance between hardware elements, with or without a barrier"
- "common external resources."

The table from ISO 26262-part9, annex C.1 illustrates the possible dependent failure causes, and also a mapping to the DFA requirements in ISO 26262-part9, clause7.4.4. According the findings, the corresponding safety measures are defined to mitigate the dependent failures.

Beyond the dependent failure analysis, freedom from interference is also employed to prove the usability of the coexistence of elements with different ASILs or QM functions. The next chapter describes the FFI and the criteria for coexistence of elements.

2.2.4 Freedom From Interference (FFI)

As mentioned before, ISO 26262 requires FFI when performing the DFA and in the case of coexistence of sub-elements with different ASIL or no ASIL (QM). For the case of coexistence of elements with different or no ASIL, the elements should be developed either with the highest ASIL of the elements or the FFI rules should ensure that the element with no ASIL or lower ASIL does not violate the higher safety related element.

According to ISO 26262-part6, annex D, FFI can be achieved by preventing the following failure modes with the proper safety measures [14]:

a "Timing and execution
  - blocking of execution,
  - deadlocks,
  - livelocks,
  - incorrect allocation of execution time, or
  - incorrect synchronization between software elements.

  EXAMPLE Mechanisms such as cyclic execution scheduling, fixed priority based scheduling, time triggered scheduling, monitoring of processor execution time, program sequence monitoring and arrival rate monitoring can be considered."

b " Memory
  - corruption of content,
  - inconsistent data (e.g. due to update during data fetch),

– stack overflow or underflow, or
– read or write access to memory allocated to another software element.
EXAMPLE Safety measures such as memory protection, parity bits, error-correcting code (ECC), cyclic redundancy check (CRC), redundant storage, restricted access to memory, static analysis of memory accessing software and static allocation can be used."

c " Exchange of information
– repetition of information,
– loss of information,
– delay of information,
– insertion of information,
– masquerade or incorrect addressing of information,
– incorrect sequence of information,
– corruption of information,
– asymmetric information sent from a sender to multiple receivers,
– information from a sender received by only a subset of the receivers, or
– blocking access to a communication channel
EXAMPLE Communication protocols can contain information such as identifiers for communication objects, keep alive messages, alive counters, sequence numbers, error detection codes and error-correcting codes."

The ISO 26262-part6, annex D provides some possible safety mechanisms to prevent, detect or mitigate these interferences.

2.3 ISO/PAS 21448 – Safety of the Intended Functionality

Safety Of The Intended Functionality (SOTIF) is defined as follows [15]:

"absence of unreasonable risk due to hazards resulting from functional insufficiencies of the intended functionality or from reasonably foreseeable misuse by persons".

As shown in the following Fig. 2.6, while ISO 26262 addresses to the systematic failures and also random hardware failures, SOTIF deals with the functional uncertainty in nominal performance.

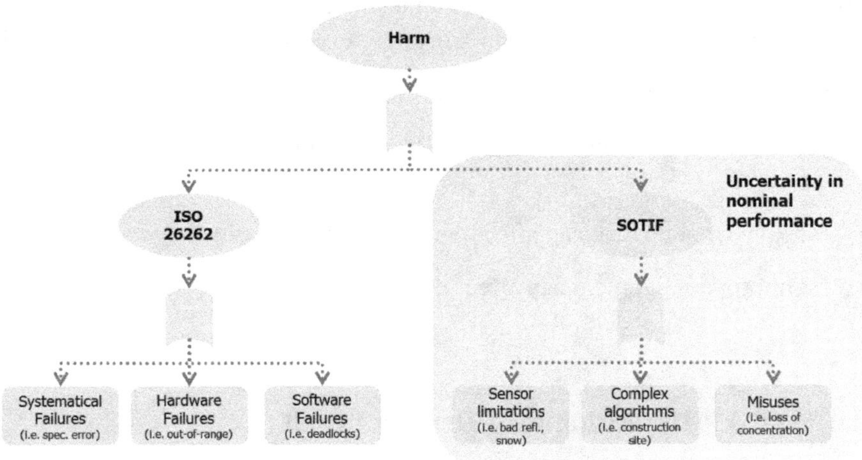

Figure 2.6: Scope of the SOTIF

SOTIF is intended to be used for the intended functionality based on the situational awareness which is derived from complex sensors and processing algorithms. This document is especially applicable for emergency intervention systems (e.g. emergency braking systems) and Advanced Driver Assistance Systems (ADAS) with levels 1 and 2 according to the OICA/SAE standard J3016 automation scales [15]. Nevertheless, the current version of SOTIF can be considered for higher levels of automation considering AD specific additional measures.

The term *triggering event* is used often in this thesis, therefore it is defined as follows [15]:

"specific conditions of a driving scenario that serve as an initiator for a subsequent system reaction possibly leading to a hazardous event".

For example, while operating on a highway, a vehicle's automated emergency braking (AEB) system can recognize a small object in front of the vehicle as an obstacle which could lead to a rear-end collision caused by a braking request due to mistaken perception.

The scope of the SOTIF is to identify and mitigate the technological limitations of the system which can lead to a potentially hazardous behavior. Some technological limitations caused by triggering events are listed as follows [15]:

- "Rain and snow can affect radar performance."
- "Rising sun in the front of the vehicle can affect the performance of a video camera."
- "A heavy woollen coat can affect the performance of ultrasonic sensors."
- "An improper alignment can affect many sensor types."

2.4 SAE J3016 - Automated Driving Levels

The number of ADAS/AD systems in road vehicles is increasing [30, 33, 34, 49]. These systems are identified and described in the SAE International standard J3016 [13] as the six levels of driving automation from "no automation" to "full automation". SAE level 2 means that the longitudinal and lateral control of the vehicle are performed by the system and the human driver is responsible for the monitoring of the environment. In contrast to level 2, the system is responsible for the perception task in SAE level 3, and the human driver is expected to intervene if this is requested by the system. The difference of SAE levels 4 and 5 to the lower SAE levels is that the systems shall be designed fail-operational to provide the fallback performance of dynamic driving tasks.

According to SAE J3016, the automation levels for road vehicles are defined as shown in 2.7:

SAE level	Name	Narrative Definition	Execution of Steering and Acceleration/ Deceleration	*Monitoring* of Driving Environment	Fallback Performance of *Dynamic Driving Task*	System Capability *(Driving Modes)*
***Human driver* monitors the driving environment**						
0	No Automation	the full-time performance by the *human driver* of all aspects of the *dynamic driving task*, even when enhanced by warning or intervention systems	Human driver	Human driver	Human driver	n/a
1	Driver Assistance	the *driving mode*-specific execution by a driver assistance system of either steering or acceleration/deceleration using information about the driving environment and with the expectation that the *human driver* perform all remaining aspects of the *dynamic driving task*	Human driver and system	Human driver	Human driver	Some driving modes
2	Partial Automation	the *driving mode*-specific execution by one or more driver assistance systems of both steering and acceleration/ deceleration using information about the driving environment and with the expectation that the *human driver* perform all remaining aspects of the *dynamic driving task*	**System**	Human driver	Human driver	Some driving modes
***Automated driving system* ("system") monitors the driving environment**						
3	Conditional Automation	the *driving mode*-specific performance by an *automated driving system* of all aspects of the dynamic driving task with the expectation that the *human driver* will respond appropriately to a *request to intervene*	System	**System**	Human driver	Some driving modes
4	High Automation	the *driving mode*-specific performance by an automated driving system of all aspects of the *dynamic driving task*, even if a *human driver* does not respond appropriately to a *request to intervene*	System	System	**System**	Some driving modes
5	Full Automation	the full-time performance by an *automated driving system* of all aspects of the *dynamic driving task* under all roadway and environmental conditions that can be managed by a *human driver*	System	System	System	**All driving modes**

Figure 2.7: Levels of Driving Automation for On-Road Vehicles [13]

2.5 Multicore Processors / Domain ECUs

The sequential computing with single core processors hits such technological limits as power wall, instruction level parallelism wall and memory wall. According to Moore's law, the number of transistors in an integrated circuit doubles proportional to the area about every two years [31, 47].

Multicore processors have become necessary in the automotive industry because of the need for more processing power, more memory and higher safety requirements. A multicore processor is a single computing component with two or more independent actual processing units (called "cores"), which are the units that read and execute program instructions.

The following figure illustrates the evolution of the processors:

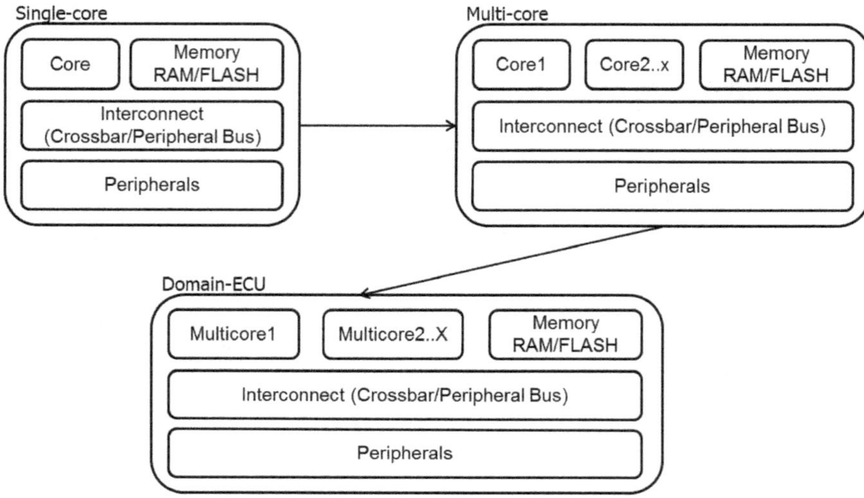

Figure 2.8: Processor evolution from single-core to domain ECUs

Domain ECU enables multiprocessor systems to be replaced with multicore processors if the content of the processors can be integrated into the cores of the multicore processors. This decreases the hardware cost, but increases common cause failures such as bus, peripherals, etc. which should be analyzed with dependent failure analysis.

In addition, the term *high performance chip* used often in chapter 3 is a type of multicore/manycore processor which consists of high performance GPUs and CPUs to execute the complex sensor data.

SoC vendors are providing additional hardware safety mechanisms with the development of multicore processor. The important hardware safety mechanisms for hardware elements are listed below:

CPU: Checker core, lockstep comparator, CRC instructions, Memory Protection Unit (MPU) for code/data/bus access
SRAM: Error Correction Code (ECC), address monitor
PFLASH: ECC for data and addresses, Error detection code (EDC) comparator, ECC monitor

Infrastructure/Peripherals: Clock monitoring, Temperature monitoring, Voltage monitoring, Watchdog monitoring, Power supply monitoring, Task sequence monitoring, Safety Management Unit (SMU)

These hardware safety mechanisms provide important advantages for decreasing the development effort of safety critical functions. For example, most multicore processors contain lockstep CPUs and a memory protection unit (MPU). The MPU could be used instead of the double storage of safety-related data required for safety critical functions classified with ASIL B, C and D. While the MPU protects the safety-related data from data corruption/overwriting, the conventional complement storage as a software safety mechanism does not only support the prevention of data corruption to achieve the freedom from interference, but also enables the detection of HW memory failures. Multicore processors provide the hardware safety mechanism "Error Correction Code (ECC)" for the detection of HW memory failures. Lockstep CPUs could be used instead of the diverse/redundant calculation of safety critical functions rated with ASIL C and D. A watchdog monitors the correct execution of safety-related functions and also the execution order of the safety-related functions using Program Flow Monitoring. Program flow monitoring and task sequence monitoring are especially important for achieving freedom from interference. In addition, SW-based fault injection tests will be provided in general from the SoC vendors and also performed for these defined HW safety mechanisms.

2.6 Architecture Description Language / EAST-ADL

The architecture can be developed using diverse modeling languages, e.g., UML, SysML or EAST-ADL (Electronics Architecture and Software Technology - Architecture Description Language). The architecture description language EAST-ADL is the modeling language that was chosen for this investigation.

The architecture description language EAST-ADL represents an ADL, which was specified initially in research project EAST-EEA (Electronic Architecture and Software Technology – Embedded Electronic Architecture) and developed further with the following research projects:

- ATESST1 (Advancing Traffic Efficiency and Safety through Software Technology) [3],
- ATESST2 [3],
- MAENAD (Model-based Analysis and Engineering of Novel Architectures for Dependable Electric Vehicles) [2],
- SAFE (Safe Automotive Software Architecture) [35] and
- Synligare [46].

EAST-ADL is a domain-specific language and is used to describe vehicle electronic systems to facilitate the development of vehicle electronics. The purpose of EAST-ADL is to offer the engineers a framework for representation and description of vehicle electronic systems in a standardized form [19]. It can be used during various development activities for the modeling of functional requirements, safety work products, and for safety analysis and design purposes [2].

This section describes the "System Model of EAST-ADL" and "Dependability and Requirements Model of EAST-ADL" in order to understand the developed approaches in chapter 4 better.

2.6.1 System Model of EAST-ADL

The metamodel of EAST-ADL consists of four different abstraction levels which fulfill specific roles. Each level considers a different phase of vehicle development, represents the complete system and provides varying perspectives of the whole EEA [19]. These four levels are the vehicle level, analysis level, design level and implementation level and they are illustrated in the following Fig. 2.9 in detail with the additional EAST-ADL extensions:

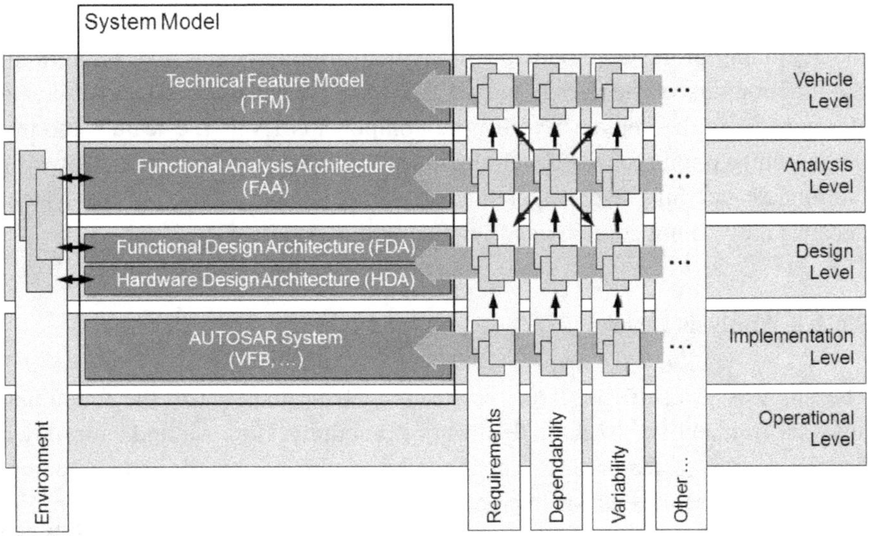

Figure 2.9: EAST-ADL overview [4]

The EAST-ADL abstraction levels are described in the following subsections.

2.6.1.1 Vehicle Level

The vehicle level enables the description of vehicle or system characteristics with many features. In this step, the question 'what' is answered in the implementation of the architecture development, but not the question "how". It makes sense to perform the hazard analysis and risk assessment on the same level simultaneously to detect and to eliminate risks as early as possible [10].

Technical feature models are created at the vehicle level to design the vehicle features [19]. Vehicle features describe, from an external perspective, vehicle components or vehicle functionalities such as the powertrain, the braking system and the steering system. These models can be created by different stakeholders and suppliers during the development process.

At the vehicle level of EAST-ADL, multiple technical feature models from the different stakeholder viewpoints can be designed in a tree structure. This allows the requirements of the different stakeholders to be considered from the beginning of the development. In the tree structure, child nodes inherit all the functionality of the parent node. EAST-ADL is constructed in a top-down structure from the vehicle level to the components level; the feature models are the entry points for the development of the system architecture. However, the models can only be considered as an abstract representation of the system because they do not contain any implementation details [10].

2.6.1.2 Analysis Level

The analysis level describes the realization of the features in the functional analysis architecture (FAA) in the form of functions. Here we find a top-down decomposition of features to functions and to abstract sensors and actuators [7]. This level represents an abstract model of the E/E system and contains the algorithmic behavior of the features in the form of abstract functions [19]. The system is modeled independently from the software and hardware. In addition, the analysis model can be used for analyzing inconsistencies and design errors in the software architecture [4].

The analysis level also helps developers to analyze the system from an engineering perspective [7]. Not only is the architecture examined, but also the interaction of the features with the real environment. The system is analyzed according to the following criteria [7]

- Completeness of requirements
- Recognition of conflicting requirements
- Identification of critical system parameters

2.6.1.3 Design Level

The design level, the third level of EAST-ADL, represents the vehicle E/E system in consideration of the functional specification and hardware architecture of the vehicle. The design level describes the system functionalities within the Functional Design Architecture (FDA) and Hardware Design Architecture

(HDA). The analysis functions in the FAA are detailed further in the FDA with regard to the functional behavior. The HDA describes the hardware architecture of the vehicle. The design level additionally enables the allocation of the functions in the FDA to the hardware elements in the HDA within an allocation diagram [8]. The FDA and HDA are described in the following subsections.

Functional Design Architecture:

As mentioned above, the design level represents the vehicle E/E system. The abstract function in the FAA is broken down into multiple functions in the FDA. The purpose of splitting FAA functions into FDA functions is to enable constraints to be met regarding non-functional properties such as allocation, efficiency or reuse. The FDA already represents the functional definition of the application software. Therefore, the FDA models can be used directly for the implementation of the functions in AUTOSAR during the implementation level of EAST-ADL [4].

The most important element of the FDA, the DesignFunction, is defined as follows:

DesignFunctionType is a specific FunctionType and is used to model the functional structure of systems at the design level. Design functions represent the logical behavior of a component and are realized by software components in AUTOSAR. At the design level, a distinction is made between the software and hardware elements. Thus, the functionality of an abstract sensor and actuator (functional devices) in the FAA is divided into three different elements, namely HardwareFunction, BasicSoftwareFunction and LocalDeviceManager. These elements are described below:

HardwareFunctionType is the interface to the real environment and represents a transfer function for the associated hardware components such as sensors, actuators, etc. A hardware function implements the logical behavior of the hardware component and provides its output in the form of electrical currents [4].

BasicSoftwareFunctionType is an abstraction of the middleware and represents a software component in the AUTOSAR BSW layer [4]. The electrical pulses from the HardwareFunctionType are processed (debouncing, filtering, etc.) by a basic software function. The outputs of the BSW function are transferred to the LocalDeviceManager as a data value.

LocalDeviceManager is functional interface to electronic devices such as sensors, actuators, etc. The LocalDeviceManager converts the electrical values of the devices, provided from the BasicSoftwareFunction, into a physical quantity [4]. For example, considering the temperature sensor, the HardwareFunction provides the sensed temperature as a voltage value. The BasicSoftwareFunction transfers this voltage value to the LocalDeviceManager, which converts the voltage value into a physical temperature value based on the sensor characteristics, e.g., temperature curves, nonlinearities, calibration, etc. [4]. The calculated temperature can be requested from other DesignFunctions. If only the LocalDeviceManager contains the characteristic of a component, the HardwareFunctionType and the BasicSoftwareFunctionType can remain unchanged by the replacement of a hardware component in the system. In such a case, only the LocalDeviceManager needs to be adapted to the characteristics of the new hardware element. Summarized, the LocalDeviceManager-DesignFunction considers the characteristics of the hardware elements as an interface to the electrical output of sensors or the electrical input of actuators and provides the physical quantity of those devices.

Hardware Design Architecture:

The HDA is used to model the hardware system architecture. The HDA describes the hardware components of the system as well as their networking in the system and also their relationship within other systems in the vehicle. The HDA library elements are Sensor, Actuator, Node and ElectricalComponent. Node represents ECUs and is connected to the sensors, actuators and other ECUs via a BusConnector. ElectricalComponent represents noncomputational hardware elements such as relays, batteries, capacitors etc. The functions of the FDA can be allocated to the hardware components of the HDA at the design level [4].

2.6.1.4 Implementation Level

The implementation level describes software architecture of the system in AUTOSAR [9]. The functions of the FDA are used as a basis for the implementation level and are implemented using AUTOSAR SW-Components. AUTOSAR is a good choice for the representation of the implementation level from EAST-ADL, because AUTOSAR is a very common software architecture standard in the automotive industry. This standardization enables the reuse of already created functions in other ECUs or in other projects. However, it takes a great deal of effort to recreate the FDA functions for the AUTOSAR application layer. Here, it would be useful to implement an extension that generates the FDA in a possible exchange format for AUTOSAR.

2.6.2 Dependability Model and Requirements Model of EAST-ADL

This subsection provides information about the EAST-ADL extensions "Dependability package" and "Requirements package" which are extended further in chapter 4. These EAST-ADL packages are further developed in a way which enables the traceability between the safety work products, requirements model and design models.

2.6.2.1 Dependability Model

The dependability model in EAST-ADL represents the system safety architecture. The dependability package contains the library elements; Item, HazardousEvent, SafetyGoal, etc., used in a model-based approach to create the safety work products according to the functional safety standard ISO 26262.

2.6.2.2 Requirements Model

The system requirements are modeled in a separate requirement diagram. The requirements model is extended and refined in iterations during the development process. The system requirements are defined in an earlier phase of the development and are detailed later in the software and hardware requirements. Therefore, the requirement package of the EAST-ADL is available on all EAST-ADL design levels. This enables the refinement of the requirements at each level of EAST-ADL. The requirements model in EAST-ADL is based on SysML and is adapted to the EAST-ADL metamodel [19]. The requirements can be configured with some specific features such as id, author, variability, etc. analog to DOORS.

The EAST-ADL requirements package allows two different types of requirements; *Requirement* and *QualityRequirement*. A Requirement enables the realization of the top level functional safety requirements, where a QualityRequirement represents the technical safety requirements on the hardware and software level. QualityRequirements can be defined as safety, availability, integrity, etc. requirements. It is also possible to connect the EAST-ADL library elements and requirements using the *Satisfy* relationship. This relationship makes it possible to find out which requirement is implemented by which functions.

3 Fail-operational Safety Architecture for ADAS/AD Systems

The self-driving technology has been developing very rapidly in the last few years. Most of the automotive companies want to launch their autonomous vehicles with SAE level 4 at the beginning of 2020, but there are still some key technological challenges to solve [28, 29]. These challenges include increasing the safety and availability through a proper fail-operational system design. In this context, this chapter describes here the possible fail-operational safety architectures for the entire ADAS/AD processing chain, from the sensors, e.g., camera, radar, lidar, etc. to the perception and decision algorithms. The solutions from this research show how the fail-operational system architecture and safety architecture can be created efficiently and how diverse redundancy can be used for ADAS/AD systems to fulfill the ASIL D safety requirements and to increase system availability.

As already described in section 2.1, the IEC61508 [6] is the basis of all industrial functional safety standards. Therefore, it is important to understand how this standard addresses the safety architecture methods. The IEC61508-part2, tables A.2-A.14 and also ISO26262-part4, annex D focus on the safety mechanisms such as input comparison/voting for E/E system with related techniques, which also contains the fail-operational safety mechanism as majority voter.

A fault-tolerant system is able to perform the specified function, even in the presence of faults [22]. This is only possible with a redundant system design. There are different ways, including passive redundancy, active redundancy and hybrid redundancy using the techniques fault masking and fault detection, to design a fail-operational system [22]. The applicability of the ISO 26262 for fail-operational automotive systems is investigated in [45] and proposals for improvement are defined for gaps found in the standard [45]. In addition, the adaptive monitoring methods for multicore processors in embedded safety-critical systems are investigated in [17].

B. Sari, *Fail-operational Safety Architecture for ADAS/AD Systems and a Model-driven Approach for Dependent Failure Analysis*, Wissenschaftliche Reihe Fahrzeugtechnik Universität Stuttgart, https://doi.org/10.1007/978-3-658-29422-9_3

There are similar examples of fail-operational architectures in avionic and rail. Avionic-related fail-operational architecture is described in [18, 32, 50]. In comparison to the avionic and road vehicle systems, rail systems are generally designed as fail-safe. The main goal in rail systems is to activate an emergency braking system in the case of a safety-critical failure. Rail applications also have a redundant system design to increase reliability [25].

The currently available papers on fail-operational automotive systems do not deal with the ADAS/AD specific fail-operational challenges. This research concentrates on solving these challenges, especially with regard to the AD processing chain and domain ECUs with multicore processors.

3.1 Introduction

The use of multicore processors in the automotive industry has become necessary in the last few years because of the need for greater processing power, more memory and added safety features. The electrification of the powertrain and autonomous driving have caused new vehicle systems to become more safety-critical. It is, therefore, necessary to investigate safety solutions, particularly for ASIL-D-Systems.

Single core processors cannot fulfill ASIL-D safety requirements because ISO 26262 requires a very high diagnostic coverage and hardware independence for the systems and allows very low hardware failure rate. To fulfill these requirements from the point of view of safety, it is necessary to either use an additional processor or multicore processors. Solutions with an additional processor are expensive, so dual core processors or multicore processors with independent cores are used at the present for ASIL-D-Systems. But the multicore processors that are currently being used will not be sufficient to fulfill the requirements in the near future because the powerful functions of the vehicle systems continue to increase and the OEMs plan to decrease the number of control units in vehicles. So the microcontroller manufacturers are developing multicore processors with more, currently up to six independent cores, which make it possible to combine different control units in a domain ECU with multicore processors.

According to ISO 26262 [14] and SOTIF [15], the safety violations that result from malfunctions of the E/E system and from random hardware failures in vehicles, and also from technological shortcomings, should be prevented or mitigated to bring the system to a safe state. This means that in the case of a fault, the system must either be switched off or be driven in a degradation mode. For SAE Level 3 and especially for SAE Level 4, innovative solutions are necessary for functional safety, system availability and system redundancy regarding fail-operational. According to SAE level 3, the driver cannot instantaneously take over control of the vehicle and also according to SAE level 4, the driver cannot be considered as a system fallback. Therefore, the system must ensure safety in case of SAE level 3 for the period of time in which the driver is still not engaged, and in case of SAE level 4 and level 5, while the driver is not available. This makes the fail-operational systems essential for autonomous driving. The system safety and system availability is very important for autonomous driving and should be increased with fail-operational architectures. This brings additional challenges for the system, software and safety architecture. For this reason, it is necessary to investigate fail-operational safety architectures for domain ECUs. The independence of the software functions allocated to the decomposition paths and main/redundant paths must be ensured in safety critical applications, which means that common cause failures, cascading failures and SOTIF-related technological shortcomings must also be safeguarded.

The purpose of this research is to provide the solutions for fail-operational safety architecture for both conventional vehicle systems and ADAS/AD systems considering the entire processing chain. Additionally, this chapter provides the methods for systematically developing a fail-operational system design and for developing a fail-operational fallback strategy in order to increase the system availability and to achieve the minimum risk condition. Furthermore, it is shown how the decomposition approach and DFA can be extended and applied for the ADAS/AD systems. Finally, the research results are illustrated within related use cases.

3.2 Safety Architecture Mechanisms

In the case of an error, if the shutdown of a system, for instance the braking system, is not possible, then it is necessary to keep the system active. Risk minimization can minimize or prevent the failure of such a system. On the one hand, risk minimization can be achieved by decreasing either the operating time or the failure rate. On the other hand, risk minimization is ensured through properly developed system safety architecture mechanisms such as fail-safe safety architectures or fail-operational safety architectures.

3.2.1 Fail-safe Safety Architecture

Fig. 3.1 shows a typical fail-safe architecture. The main principle is that the system, with its main components; sensor, electronic control unit and actuator, is monitored by various diagnostic functions and if a failure occurs, the system is switched off.

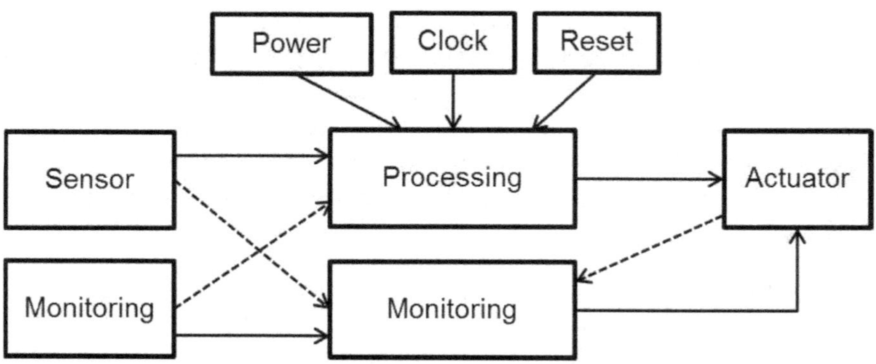

Figure 3.1: Fail-safe safety architecture

3.2.2 Fail-operational Safety Architecture

As mentioned before, it is not always possible to switch off the system in the case of a failure. The system architecture requirements for fail-operational safety architecture can be implemented as diverse redundancy or homogenous redundancy. Diversity is achieved with two or more different hardware and software applications, which are developed by different companies or teams, or which are based on diverse specifications. In a homogenous redundant system, the hardware, e.g. processor, is developed by the same company and same development teams. The software is produced by two or more equal instances. In the following subsection, some of the existing fail-operational systems are described.

3.2.2.1 1-out-of-2 Safety Architecture (1oo2):

This safety architecture consists of two independent processing units. These processing units are able to control the actuators independently. If one processing unit fails, the system is still operational. It is advantageous to apply this safety architecture in conjunction with a diagnosis. Fig. 3.2 shows the typical 1-out-of-2 fail-operational safety architecture [6].

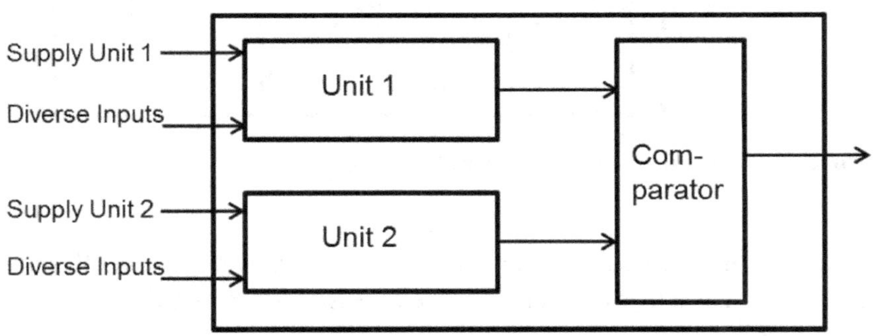

Figure 3.2: Fail-operational safety architecture (1oo2)

3.2.2.2 2-out-of-3 Safety Architecture:

The 2-out-of-3 safety architecture in Fig. 3.3 is a well-known fail-operational architecture. This approach is also known as Triple Modular Redundancy (TMR) and is widely used in the avionic field as a reference architecture for safety-critical fail-operational systems [6]. In this approach, three units are calculated redundantly and also independently from each other. It is very important to use the different input signals and power supplies in the function calculations. The calculation results are compared and if at least two of the calculated units have the same result, the output is true. If one of the units fails, the system can continue with the remaining two units. So it is still possible to compare these units regarding the correctness and the system is still fail-operational.

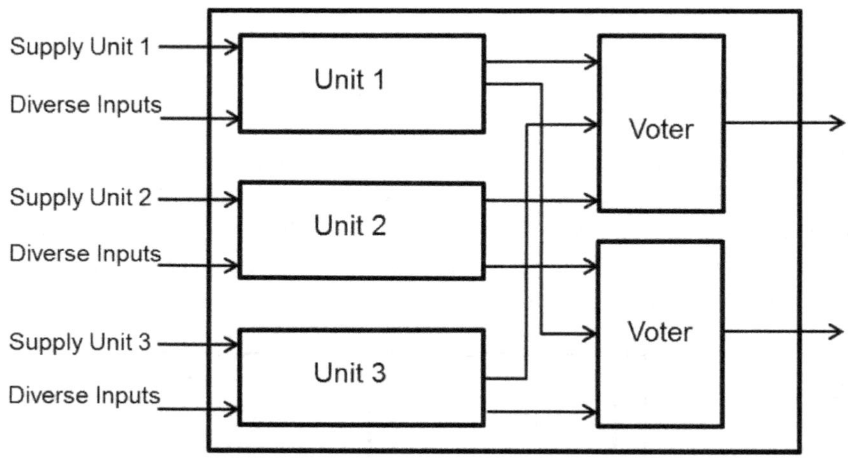

Figure 3.3: Fail-operational safety architecture (2oo3)

This approach is very suitable for a fail-operational architecture, but there are also some disadvantages, for instance the necessity for more ECUs, and more power consumption.

3.2.2.3 2-out-of-2 Safety Architecture:

In comparison to the 2oo3 architecture, the realization of a 2oo2 system is easier, as shown in 3.4 [6]. If one of the two components fails in one of the two channels, the system continues to be available with the one remaining channel. But in this case, the system is no longer fail-operational. It is not good to run a safety-critical system with only one channel, but it is possible for a certain period of time.

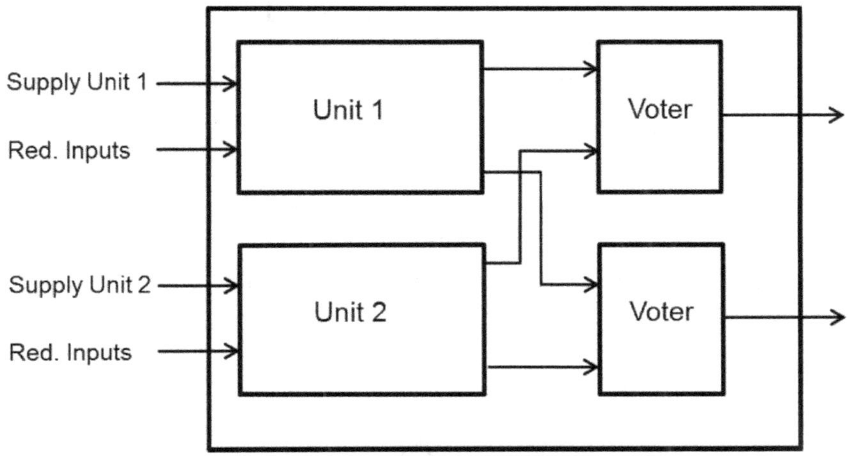

Figure 3.4: Fail-operational safety architecture (2oo2)

3.2.2.4 2-out-of-2 PD Safety Architecture:

The technical paper [27] describes the 2-out-of-2 (2oo2) DFS architecture, which consists of two either equal or diverse subsystems. In this architecture concept, each subsystem uses its own monitoring concept and is designed as a fail-safe system to detect its own errors (Fail-Safe (FS)). The disadvantage of 2oo2 DFS is that unnecessary processing power is spent because of the cyclic monitoring of units. The Fig. 3.5 shows the newly developed fail-operational safety architecture 2-out-of-2 Performance Diagnosis (2oo2 PD). In this approach, the results from the redundant or diversity redundant units are compared to each other. If the results are equal, then these results are used directly to control the actuators. If the results are not equal, then the units are checked by the monitoring functions to detect and isolate the defective unit. The faulty unit is disabled and the system is controlled by the correctly functioning unit, thus the system remains fail-operational.

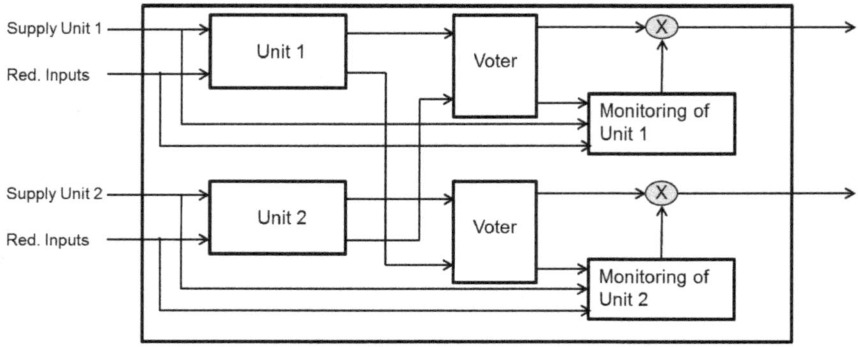

Figure 3.5: 2-out-of-2 Performance Diagnosis (2oo2PD)

3.3 Fail-operational Safety Architecture for Conventional Systems

If a system deactivation could potentially lead to hazardous behavior, i.e. one or more safety goals are violated by the shutdown, then it is necessary to develop fault-tolerant safety mechanisms that at least enable emergency operation in the case of a failure. These systems are called fail-operational systems and are investigated on the microcontroller level and ECU level in the relevant works [26, 27]. In contrast, the approach described in this chapter shows the fail-operational safety architecture considering domain ECUs developed for conventional systems such as powertrain and steering.

Typical fail-operational systems are 2oo3 systems, which, as mentioned before, are widely used in the field of avionics. The redundant architecture structure of such systems provides very high availability. 2oo3 systems consist of three fully independent redundant elements from the sensor to the actuator. These three different paths are checked against each other for plausibility to find and isolate the defective path in the case of a failure. If a failure occurs, the system can be maintained with the two other remaining systems.

At the same time that the E-mobility and autonomous driving technologies are causing the use of multicore processors to increase, OEMs and system providers seek to reduce the number of ECUs in the vehicle. In this context, this research investigates whether fault-tolerant systems such as the 2oo3 safety architecture can be realized with domain ECUs within multicore processors. Fig. 3.6 shows the solution that was developed for the fail-operational architecture. The main feature of this approach is the usage of the second processor as a backup for the faulty processor or arithmetic core. The safety-critical functions are calculated redundantly in each processor and the results are compared with each other. If the results are not equal, then, to find the fault path, these results are compared with the results from the other processor. The system can then actively work with the remaining correctly functioning paths.

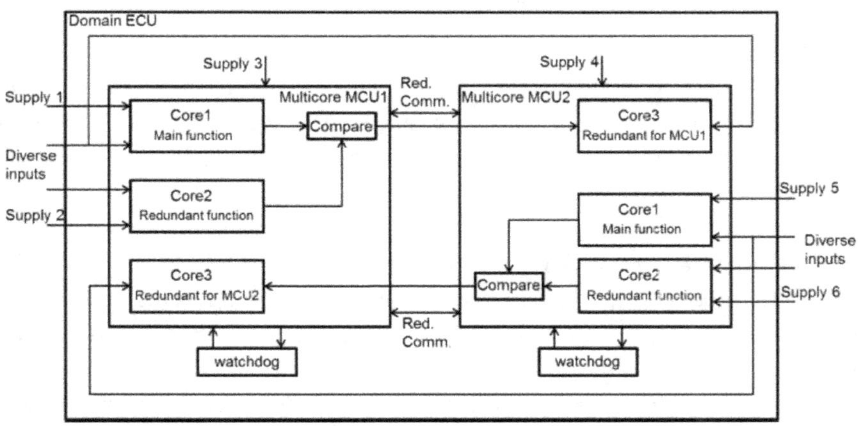

Figure 3.6: Fail-operational safety architecture considering domain ECUs within multicore processors

The benefit of this approach is the increase in system availability provided by the fail-operational architecture within the redundant cores. It is highly recommended to use two different multicore or manycore processors from different suppliers in order to decrease the systematic failures.

3.4 Fail-Operational Safety Architectures for ADAS/AD Systems

As explained above, fail-operational systems are essential for autonomous vehicles. SAE level 3 vehicles are only required to be fail-operational for a short time until the driver can react to a request to take over control of the vehicle. Vehicles with SAE level 4 and level 5 must be safe and fail-operational during the entire driving cycle to be acceptable to customers. If the vehicle systems are deactivated due to a system failure, it is still possible that the vehicle is safe in an one-lane construction zone or in a tunnel at a limited velocity. But because availability is an important factor for the customer acceptance of fully autonomous vehicles, autonomous vehicles must be designed to be failure tolerant, namely with diverse redundancy from the sensor to the actuators.

Fig. 3.7 shows the general processing chain for autonomous vehicles. ADAS/AD systems can be thought of as humans. The eyes are the sensors, the brain and neurons are the high performance ECUs and the actuators are the hands and feet. The process starts with sensing in order to detect pedestrians and the environment, both static and dynamic objects. The currently-used sensors for autonomous vehicles are cameras, radars, lidars, ultrasound sensors and DGPS. It is important to have a 360° field of view for the vehicle to collect all the information that is important for safe autonomous driving. This data is then used for the sensor fusion with an appropriate filter, for example, a Kalman filter or a Bayesian filter, to detect and classify objects and the environment and also to predict the subsequent scenarios. In the next step, the path planning is performed. After the situation is analyzed and interpreted, decisions are made about emergency braking, emergency steering and also normal driving function. According to these decisions, and depending on the vehicle model, the actuators are controlled by either the domain ECUs or by an additional ECU with a safety microcontroller.

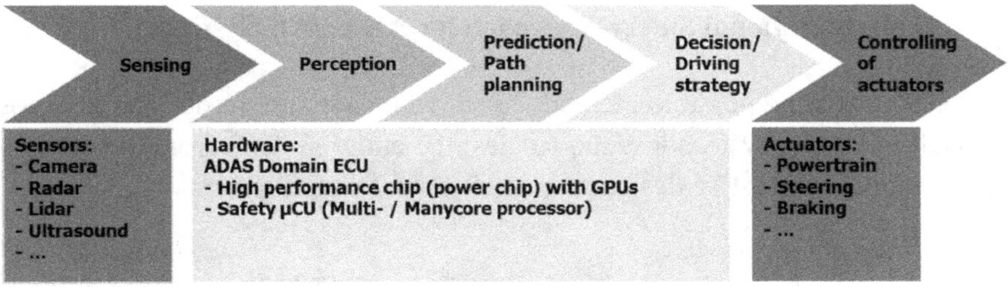

Figure 3.7: General processing chain of ADAS/AD systems

To keep the system fail-operational, it is important to detect all possible causes for failure in the processing chain and to mitigate these causes. The conventional hardware failure modes are defined in part 5 of the ISO 26262. The safety mechanisms for these conventional failures are already implemented for ADAS/AD, but they are not sufficient for a safe vehicle. The hardware metrics of the systems can be achieved by following the parameters described in the semiconductor safety manual. In addition to the existing safety measures, the SOTIF measures for improving technological shortcomings to increase the nominal performance should be developed and addressed according to ISO/PAS 21448 (SOTIF) as shown in Fig. 3.8.

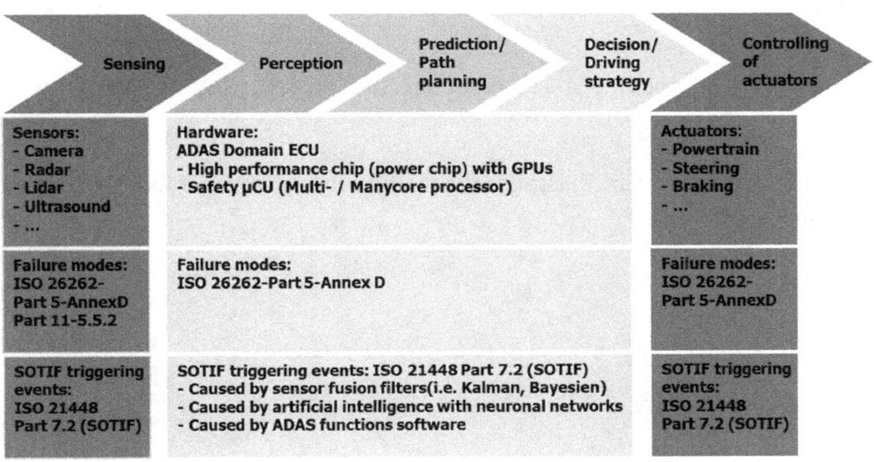

Figure 3.8: Failure modes of ADAS/AD systems

3.4.1 Fail-operational Safety Approach for ADAS/AD Systems

Fig. 3.9 shows the possible methods for a redundant system design. It is not enough to have only sensor redundancies; the entire system, from the sensors to the actuators, must be designed in a fail-operational manner.

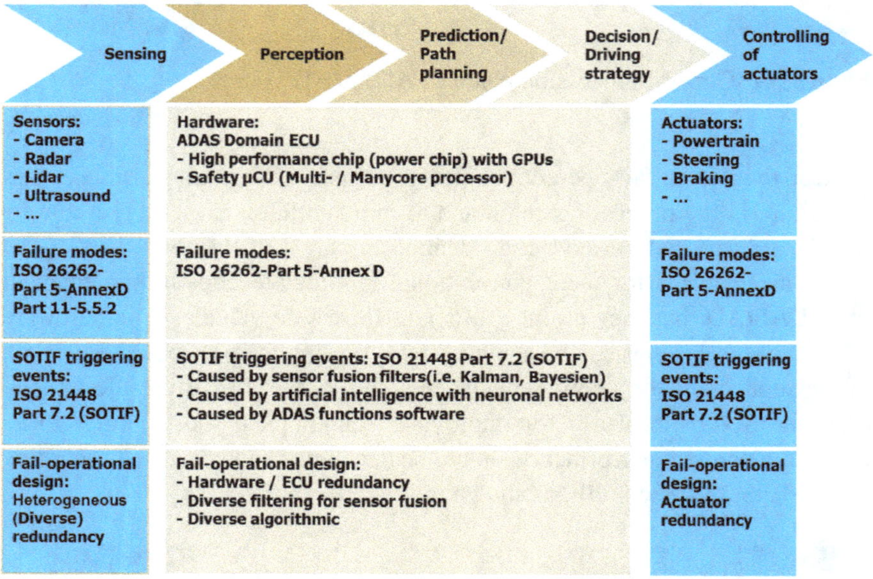

Figure 3.9: Fail-operational safety architecture for ADAS/AD systems

The following flowchart illustrates the method for achieving a fail-operational system.

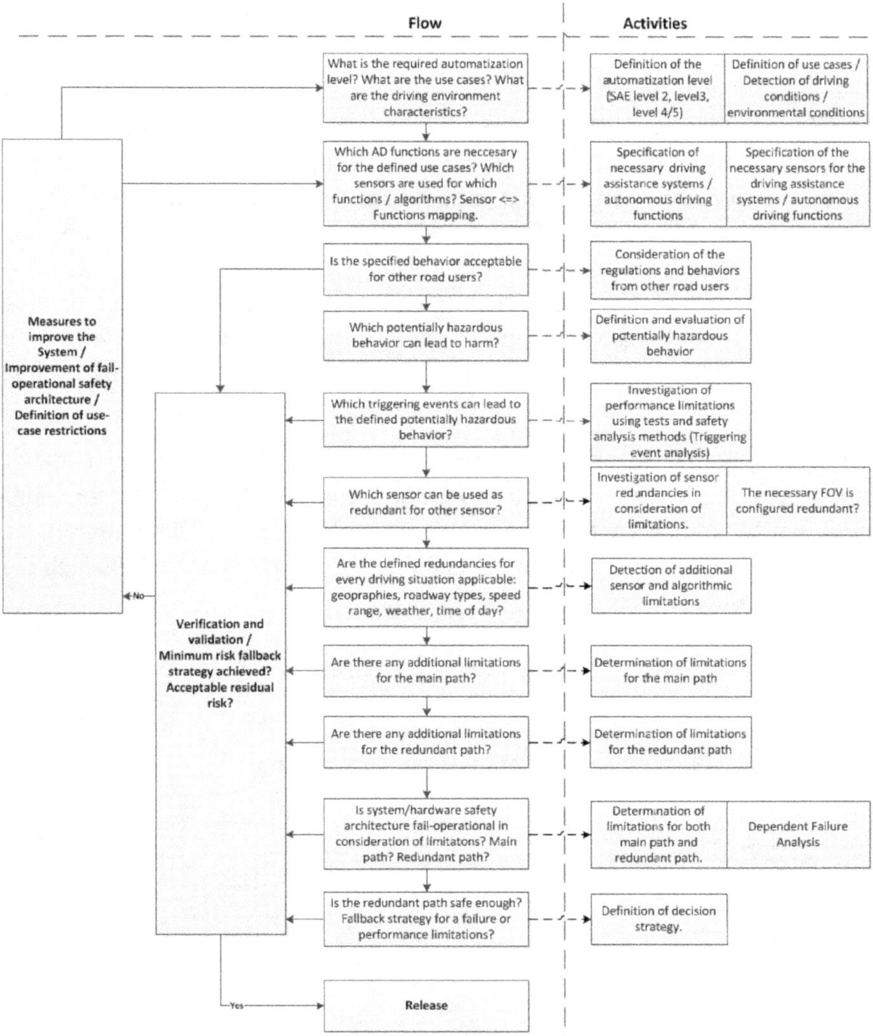

Figure 3.10: Method for the development of fail-operational safety architecture

The parts of the flowchart are described in the following sections.

3.4.1.1 Sensor Redundancy / Mapping of Functions to Sensors:

In order to stay fail-operational in every driving situation and under all weather conditions, it is necessary to thoroughly investigate the sensor characteristics and limitations. It is not enough to only consider the field of view characteristic for the mapping between the sensors and the autonomous driving functions, because there are other external factors such as weather conditions and infrastructure which can affect the sensor performance negatively. In such a case, the system is no longer fail-operational, unless these external factors are already considered by the system design and the system has redundancies regarding these cases. The goal of SOTIF is to minimize such unknown scenarios and reduce any negative effects to an acceptable level.

The following Tab. 3.1 shows a proposal for a possible diverse redundancy mapping of the sensors to the function. The focus is not to prove the completeness, but to show an approach for efficient fail-operational designing of the system. The mapping can differ depending on the conditions under which the product is used.

Table 3.1: Mapping ADAS functions with sensors

	Radar			Lidar	Camera		Ultrasound	DGPS	IMU
	Long Range	Mid Range	Short Range		60° FOV	120° FOV			
Adaptive Cruise Control (ACC)	X		X	X					
Automatic Emergency Braking (AEB)	X		X		X				
Lane Keeping Assist (LKA)				X	X				
Lane Centering (LC)				X	X				
Lane Departure Warning (LDW)					X	X			
Traffic Jam Assist (TJA)	X		X	X	X	X			
Collision Avoidance	X			X	X				
Traffic Sign Recognition					X	X			
Park Assist							X		
Valet Parking			X	X	X	X			
Overtaking / Lane-Passing	X		X	X	X	X			
Highway Assist	X	X	X	X	X	X			
Cross Traffic Alert		X	X						
Surround View					X	X			
Blind Spot Detection		X	X						
Rear Collision Warning		X	X						
Localization								X	X

In addition, the following Fig. 3.11 shows a proposal for a possible diverse redundancy mapping of the sensors to the functions for automated driving systems from SAE level 3 to SAE level 5.

SENSORS / ADAS-FUNCTIONS	Radar			Lidar	Camera		Ultrasound	DGPS
	Long-Range	Mid-Range	Short-Range	Lidar	60° FOV	120° FOV		
Longitudinal control functions	X			X	X		-	-
Lateral control functions	X			X	X		-	-
Collision avoidance functions	X			X	X		-	-
Valet parking function	X			X	X		X	-
Localization	-			-	-		-	X

Figure 3.11: Mapping AD functions with sensors

3.4.1.2 Electronic Control Unit Redundancy / HW Redundancy:

After the sensor data is collected, the data should be calculated and analyzed within a high performance domain electronic control unit. The processing power and the memory of conventional ECUs with multicore processors is not sufficient for the ADAS/AD function calculations above SAE level 3 because of the large amount of data from the various complex sensors and because of the machine learning algorithms. To perform the calculations for fully autonomous vehicles, the high performance chips are necessary and there are already some high performance ECUs on the market. AD Domain ECUs consist of one high performance chip and one conventional safety multicore processor. In general, the high performance chip is used for the processing of complex sensor data and for the execution of complex perception and decision algorithms. The conventional multicore processor is used to control the actuators according to the information from the high performance chip. This research illustrates in the following text, a possible fail-operational hardware safety architecture depending on the use cases. The fail-operational safety architecture for domain ECUs depends on the availability requirements, i.e. the requirement to keep the system fail-operational for only a very short time with driver interaction at SAE level 3, is different than ensuring that the system is fail-operational for the entire driving cycle without the driver at SAE level 4 and 5. As mentioned before, at SAE level 3, if the system prompts the driver, he should be able to assume control of the vehicle. For the SAE levels 4 and 5, if a system failure occurs, fully autonomous vehicles must be able to attain a safe state. Therefore, the system fallback concepts for vehicles with SAE level 0-3 are different from the fallback concepts for vehicles with SAE level 4 and level 5. Vehicles with SAE level 4 and 5 must be equipped with fail-operational redundant systems which can be used as fallback systems. For these systems, it is necessary to have at least a 2oo2D hardware safety architecture or, even better, a 2oo3 hardware safety architecture, namely triple modularity, because of reliability and availability requirements.

The safety architectures shown in Fig. 3.12, Fig. 3.13 and Fig. 3.14 are designed within the following conditions:

- It is important that the information from the various sensors is processed independently.
- A high performance chip calculates the functions redundantly and provides two independent signals to the multicore processor/safety chip (illustrated as Multicore MCU in Figs.).
- The information signals to the safety chip must be transmitted via redundant safety communication without any corruption.
- The majority selection is performed in parallel on two different cores.
- The high performance chip, the individual cores of the safety chip and the different channels of the performance chip must be supplied with redundant supply voltage.
- Watchdogs must monitor the individual chips to detect if they are running properly.

The Fig. 3.12 shows the possible fail-operational safety architecture with respect to AD domain ECUs. The current AD domain ECUs generally consist of one high performance chip within GPUs and CPUs and one conventional safety multicore or manycore processor. The process starts with the sensor signals being independently monitored according to ISO 26262-5 for sensor failure modes such as out-of-range, offsets, oscillations etc. The sensor signals are also monitored for sensor characteristics such as EMC disturbance, accuracy, etc. from ISO/PAS 21448, clause7 (SOTIF). The correct sensor data is then used to redundantly implement sensor fusion within different GPUs in order to detect and classify the objects. Next, behavior prediction is performed to find out the intent of each object on the road. In the following step, the path planning is done independently using the distinct information. Decision making is performed either in the high performance chip (illustrated as power chip in Figs.) or in the safety processor. This architecture is suitable only up to SAE level 3 because of the limited redundancy. If the power chip fails, the multicore safety chip is only used as a fallback for important safety-critical functions of the power chip. New manycore processors also have radar interfaces, so these features can be used for redundancy to bring the vehicle to a safe state in the case of a power chip failure.

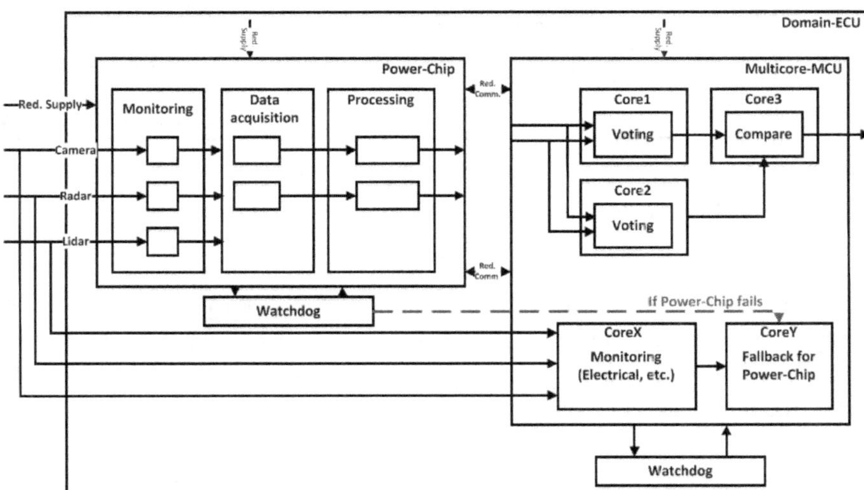

Figure 3.12: Fail-operational safety architecture up to SAE level3

The Fig. 3.13 illustrates the 2oo2D architecture in the domain ECU context. In comparison to the previous safety architecture, the second high performance chip (Power-Chip2) calculates the functions and algorithms with diverse redundancy. The results are compared to the safety chip results to find out whether they are equal. In addition, each high performance chip has self-diagnostic capability, so it is possible to detect the fault path if the results are not equal. In this case, the system remains fail-operational by use of the correct path. In the case of unequal results, if it is not possible to determine the false path, then the system is no longer fail-operational and must be shut down to achieve a safe state.

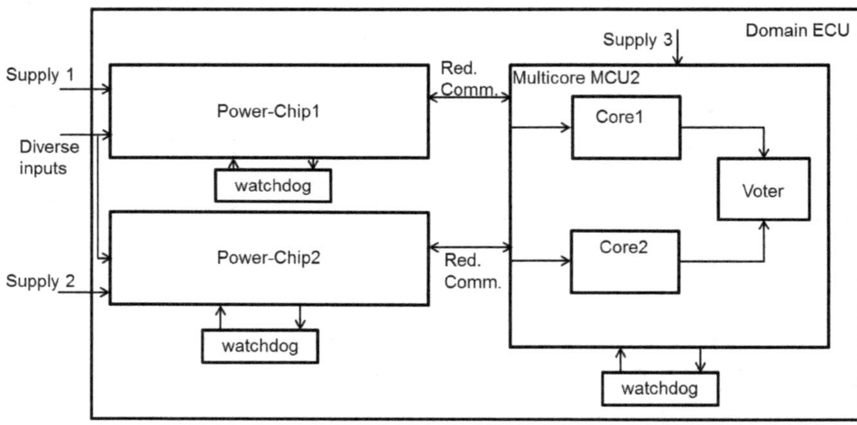

Figure 3.13: Fail-operational safety architecture: 2oo2D

The Fig. 3.14 shows the concept of triple modularity on the ECU level. In comparison to the 2oo2D safety architecture, this architecture provides greater availability. The three independent paths are compared within the voters to detect the fault path. The system is fail-operational until 2 out of 3 ECUs fail.

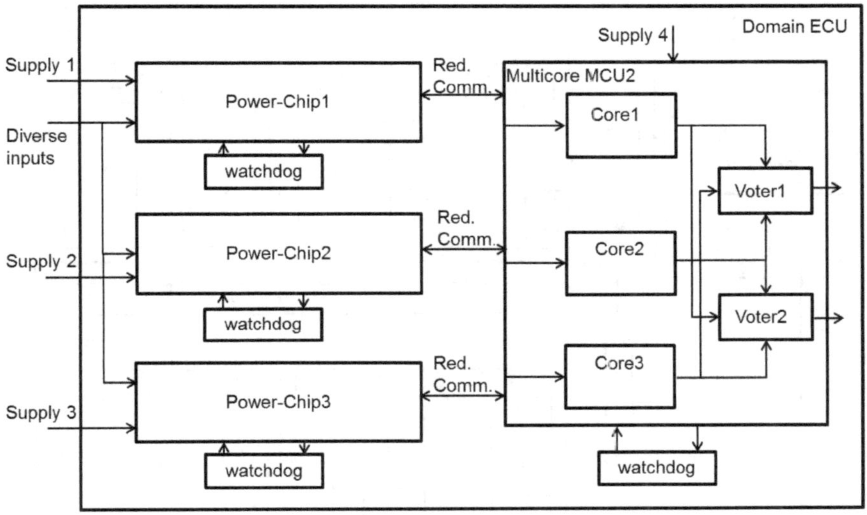

Figure 3.14: Fail-operational safety architecture: 2oo3

The Fig. 3.15 illustrates the fail-operational processing chain of ADAS functions and algorithms within a high performance chip. The high performance chips on the market consist of high performance GPUs and CPUs, so they are suitable for designing fail-operational safety architecture. The sensor signals are first diagnosed regarding electrical failure, e.g., short to ground, and also regarding sensor characteristics, e.g., loss of pixel data. Then the functions and algorithms for perception, path planning and driving strategy are performed diversely within redundant GPUs. The results of the functions are compared in a CPU, ensuring diverse redundancy. Finally, the outputs of the high performance chip are compared in the independent safety chip as shown in Fig. 3.12, Fig. 3.13 and Fig. 3.14.

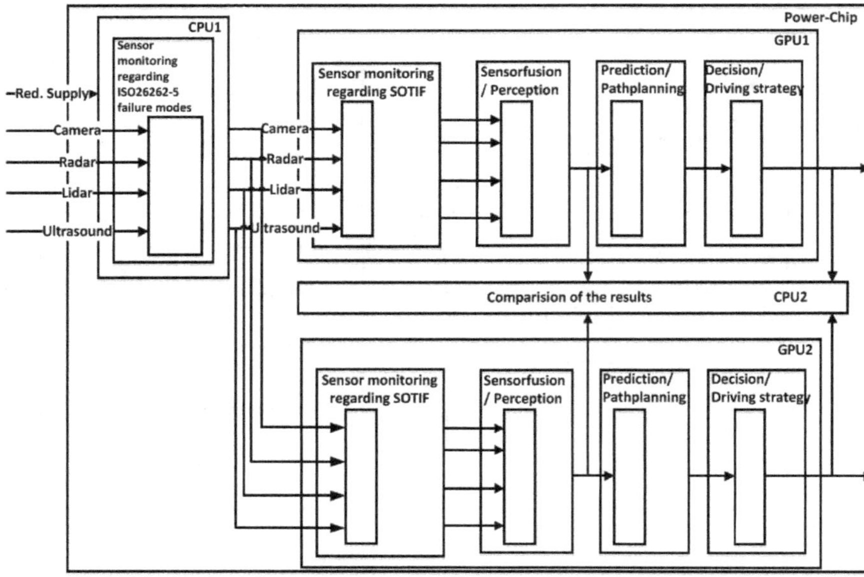

Figure 3.15: Illustration of processing chain of ADAS/AD within high performance chip

If the sensors or the electronic control units are no longer redundant, or if, in general, the processing chain path from sensors to actuators is no longer redundant, then the vehicle must be put into a safe state. Alternatively, the reason for the redundancy failure can be evaluated and checked with continued processing using only one path to enable further driving availability under certain conditions such as velocity limitation, restricted environment, etc.

At this point, the questions arise as to which of the illustrated safety architectures is most suitable or which of the safety architectures is more suitable under specific conditions. The answer to these questions depends on the system requirements. Simplified, the architecture in Fig. 3.12 is suitable for the systems up to SAE level 3, because the availability requirement is lower and the driver can be considered as a system fallback. The safety architectures Fig. 3.13 and Fig. 3.14 can be suitable for fully self-driving vehicles with SAE level 4 and level 5, because both safety and availability are very important for such systems. If the availability requirement is lower, e.g., if the vehicle is only used in a restricted area such as an airport; then the safety architecture in Fig. 3.13 can be selected. There are many other factors, including use cases, costs, etc., to consider when choosing the appropriate safety architecture. The decision depends greatly on how the item definition and use cases for the vehicle are defined. It certainly makes a difference whether the vehicle is in operation only during the day and not at night, or only in good weather and not in the rain and snow and, of course, whether the vehicle is in operation only in a restricted area or also on the highway or in an urban environment with many passengers. If there are no cost requirements, then, similar to avionic field, the safety architecture in Fig. 3.14 would be advisable. So far, the automotive industry has been cost-driven on such issues. However, there should be a change in attitude, especially for fully self-driving vehicles, so that these vehicles can be designed in a much safer way to achieve the minimum risk condition in every driving situation and to increase the availability and also to gain the acceptance of the customer for fully self-driving vehicles. The systematic approach for the development of fail-operational architecture in Fig. 3.10 can also be taken into account when selecting the HW safety architecture.

The next subsection describes an intelligent fail-operational fallback strategy to increase the system availability and to achieve the minimum risk condition.

3.4.1.3 Intelligent Fail-operational Fallback Strategy to Achieve Minimum Risk Condition

It is very important to develop a suitable fallback strategy to achieve the minimum risk condition in case of a system failure or system limitations. In the development of this fallback strategy, there are two important aspects to consider, safety and availability. System safety is essential for AD systems, but availability also plays a significant role in fulfilling customer expectations. Therefore, these two facets are considered in the following generic intelligent fallback strategy. This fallback strategy is developed for a triple modular system, but it is also applicable in the same way for Xoo2 systems.

As shown in the following flowchart, the system consists of three computation paths. The redundant paths shall be designed independently in order to increase the availability. The first path and second path contain two operating modes in order to increase the availability. The first operating mode is the normal operating mode and is active until an error occurs. There are several types of errors in the system and they should be considered differently because of their differing system impacts. In case of a hardware failure (ECU or μC) or sensor failure, which is very essential for the safety function and there are no further sensor redundancy for this information, the fail-operational system switches directly to second path. In the case of SOTIF-related technological shortcomings such as sensor limitations due to weather conditions, or sensor failures with existing redundant sensor information, the first computation remains with a degradation mode, for example, in heavy rain, the vehicle continues to drive at a reduced velocity. If these technological deficiencies no longer affect the system operation, then the system goes back to the normal operating mode. If an additional hardware failure, or a sensor failure without a redundant sensor, or any additional SOTIF-related technological limitations that affect the remaining sensors and algorithms occur in the degradation mode, then the system switches to the second computation path. This procedure is performed in the same way for the second path until a switch to the third path. The third path is designed to achieve the minimum risk condition with a limp aside mode.

The Fig. 3.16 illustrates the intelligent fail-operational fallback strategy for minimum risk condition.

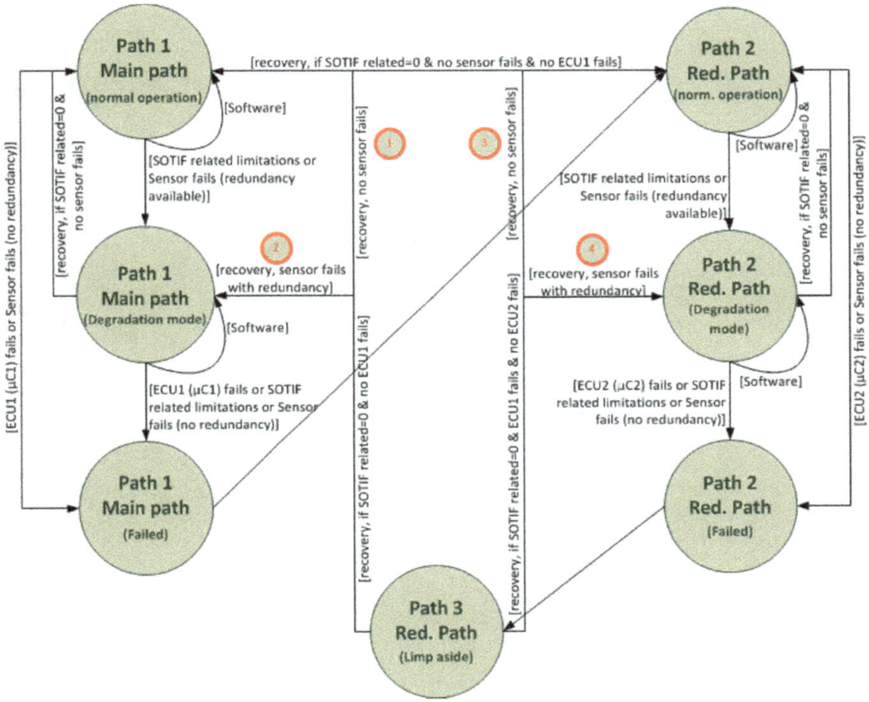

Figure 3.16: Fail-operational fallback strategy

The Fig. 3.17 shows the fail-operational fallback flow for ECU failures and also sensor failures, where no redundant sensor information for this sensor exists. As illustrated below, there is no possibility to recover the system in the case of a hardware failure, so it is absolutely essential to design the computation paths independently and diversely redundant to prevent the simultaneous failure of both the ECUs (μCs) and the sensors due to common cause failures such as supply voltage, high temperature, or systematic failure.

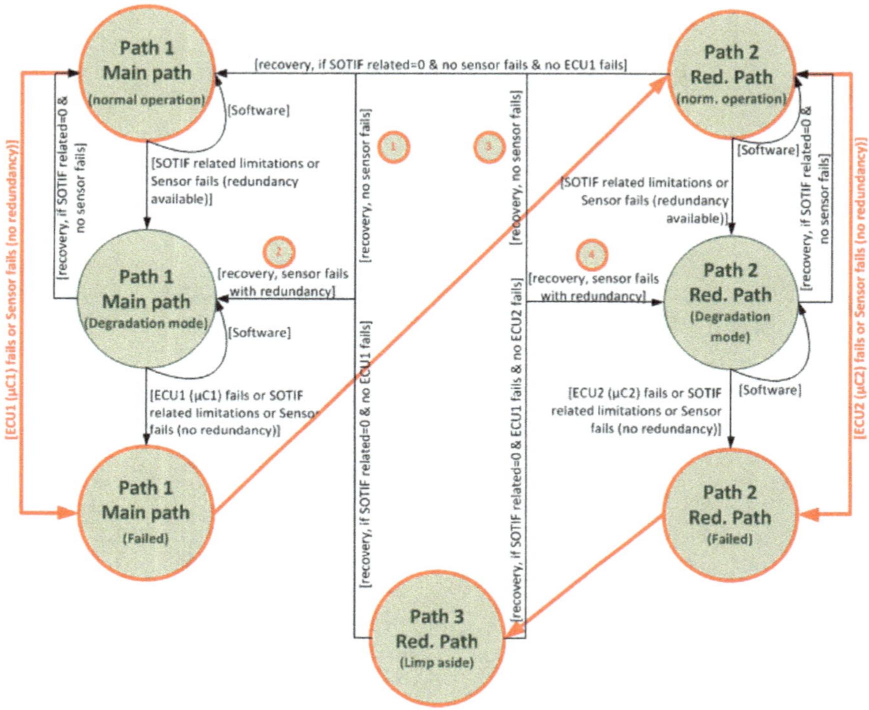

Figure 3.17: Fail-operational fallback strategy for ECU and sensor failures

Fig. 3.18 shows the intelligent fail-operational fallback flow for SOTIF-related limitations and also sensor failures, where additional redundant information for this sensor exists. It is possible, that the first and second computation paths are affected in this failure combination, sensor failures and/or SOTIF related technological shortcomings. In such a case, the third path brings the vehicle to a safe state. In this case, the system checks whether the failures are still there or not. If the SOTIF-related technological shortcomings are no longer available, then the system switches to the following operating modes:

1 First path normal operating mode, if no ECU failure and no sensor failures

2 First path degradation mode, if no ECU failure and sensor failures with redundancy

3 Second path normal operating mode, if ECU1 fails, but no ECU2 failure and no sensor failures

4 Second path degradation mode, if ECU1 fails, but no ECU2 failure and sensor failures with redundancy

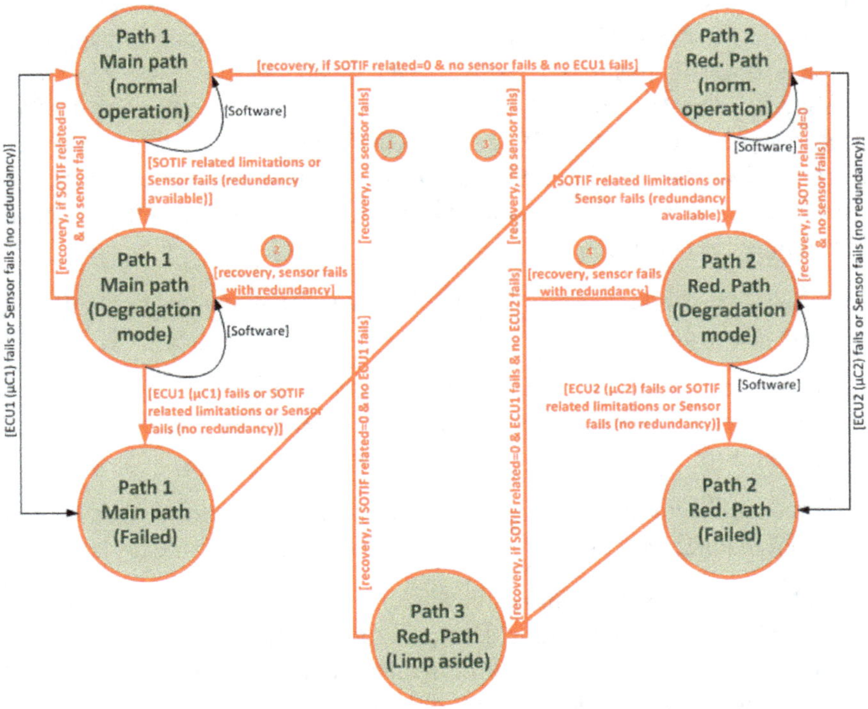

Figure 3.18: Fail-operational fallback strategy for SOTIF-related limitations and sensor failures

3.4.2 ASIL Decomposition for ADAS/AD Systems

3.4.2.1 ASIL Decomposition in General

That new vehicle systems have become more safety-critical through the electrification of the powertrain and due to autonomous driving, has already been mentioned. The ISO 26262 standard provides for the possibility to apply the decomposition approach to the development of safety critical systems, particularly ASIL-D rated safety systems. An appropriate decomposition has the advantage of reducing the ASIL rating of the top level safety requirement derived from safety goal, but ASIL decomposition requires the redundancy of safety requirements, which should also be allocated to sufficiently independent architectural elements.

ISO 26262-part9, clause5 mentions the following requirements to the decomposition approach [14]:

- "As a basic rule, the application of ASIL decomposition requires redundancy of safety requirements allocated to architectural elements that are sufficiently independent."
- "If the architectural elements are not sufficiently independent, then the redundant requirements and the architectural elements inherit the initial ASIL."
- "In the case of use of homogenous redundancy (e.g. by duplicated device or duplicated software) and with respect to systematic failures of hardware and software, the ASIL cannot be reduced unless an analysis of dependent failures provides evidence that sufficient independence exists or that the potential common causes lead to a safe state. Therefore, homogenous redundancy is in general not sufficient for reducing ASIL due to the lack of independence between the elements."

The developed fail-operational architecture for conventional systems is also suitable for applying the decomposition approach to ASIL D systems according to ISO 26262-part9 as shown in Fig. 3.19, because the domain ECUs in this concept offer the required sufficiently independent architectural elements.

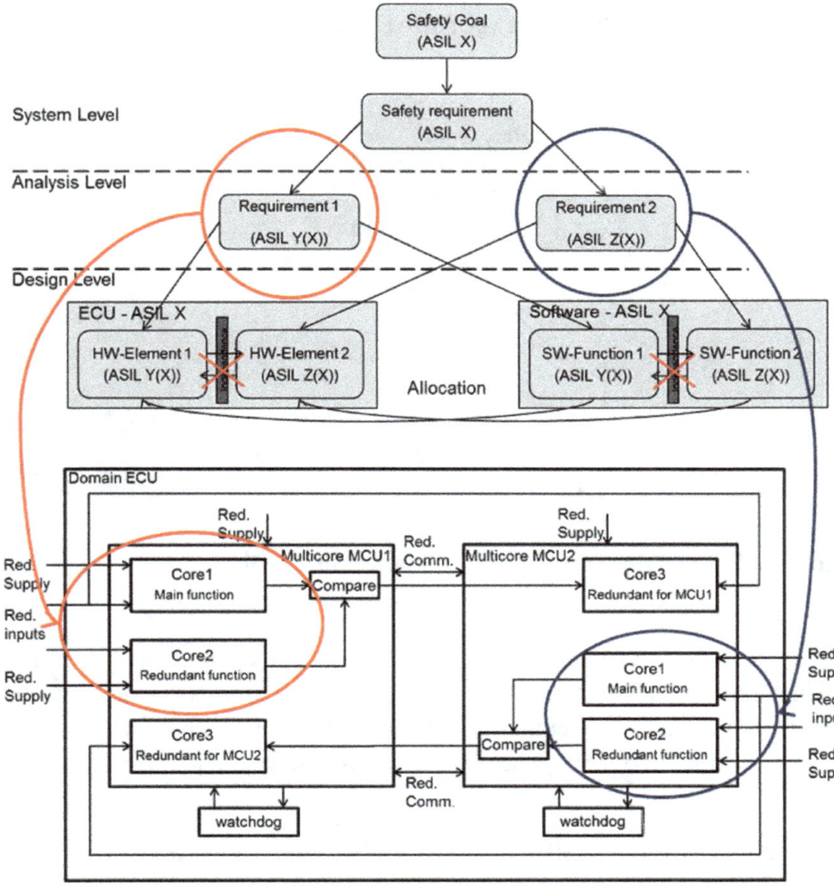

Figure 3.19: Independent distribution of decomposed parts

3.4.2.2 ASIL Decomposition for ADAS/AD Systems

Safety goals for SAE level 3 and level 4 functions, e.g., longitudinal control, lateral control and collision avoidance, are generally classified with ASIL D, depending on the related scenarios. In order to achieve the ASIL D safety goals, the AD processing chain from sensing the environment and traffic participants to the decision making must be implemented within ASIL D. An ASIL

D safety goal impacts the entire ADAS/AD processing chain as shown below in Fig. 3.20.

Figure 3.20: Safety goal impact on AD processing chain

This means that the sensing of the environment, perception, path planning, decision making and the controlling of the actuators must also be developed in ASIL D. In the following text, this is illustrated with an example which is also detailed in subsection 3.5.3, where some possible safety requirements are also listed.

Malfunction: Missing collision avoidance behavior. The system fails to activate collision avoidance functions when it should. For example, the vehicle does not stop for pedestrian or red traffic light.

Safety goal: Collision avoidance shall be ensured (ASIL D)

Safety requirements: The ADAS/AD function shall ensure the correct control of the actuators, based on the input, to avoid collisions with traffic participants, static obstacles, and dynamic obstacles (ASIL D).

To fulfill this requirement, the ADAS/AD processing chain must be implemented in ASIL D. For example, the ADAS/AD function shall correctly detect and characterize traffic participants and the environment including static and dynamic objects.

The algorithms and corresponding hardware must also be also developed in ASIL D to fulfill this requirement. But the hardware, from the sensors to the high performance chips (graphic processors) where the ADAS/AD algorithms are implemented, is not sufficient to achieve ASIL D. This is why it is necessary to use decomposition to implement ASIL D safety goals for ADAS/AD systems, but the use of decomposition creates some additional challenges for ADAS/AD systems in comparison to conventional systems. The following text illustrates the decomposition concepts for ADAS/AD systems which have been developed in this research.

This approach, showing how decomposition can be used for ADAS/AD systems, is based on the algorithms being performed within sufficiently independent hardware elements, i.e., high performance chips. The independent channels use independent sensor elements because sensors with an ASIL D rating are not on the market. During decomposition, it is important to achieve a minimum risk condition independently for each decomposition path. For conventional systems, the minimum risk condition is fail-safe, which means the system is deactivated so that the vehicle enters a fail-safe state. For ADAS/AD systems, it is important to achieve a fail-operational state. In the case of a failure in one decomposition path, the second decomposition path is used to switch the system to the redundant path. This redundant path is then responsible for further safe driving or used to bring the vehicle to a minimum risk condition.

In the following approach illustrated in Fig. 3.21, the results of the independent decomposition paths are compared to each other. In case of an inequality, the second decomposition path switches to the redundant path.

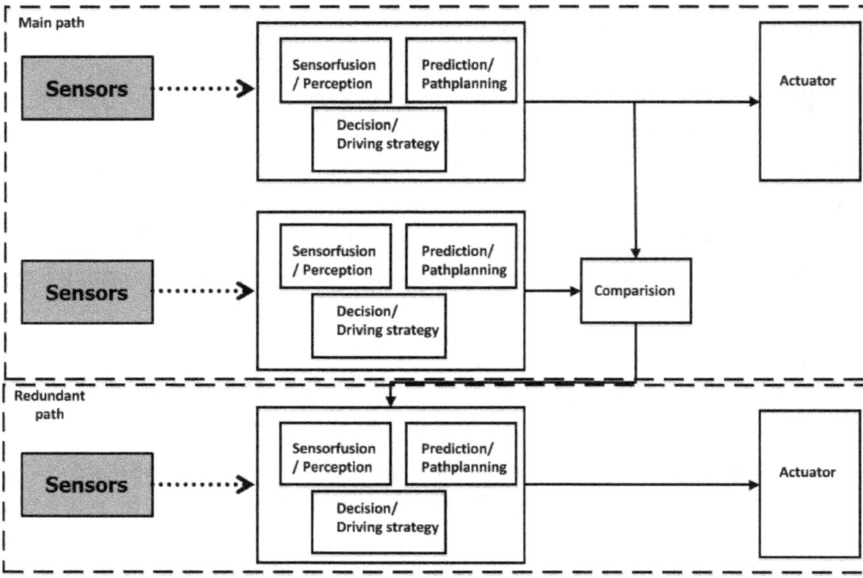

Figure 3.21: First approach: ASIL decomposition for ADAS/AD systems

As shown in Fig. 3.22, in this approach it is necessary to use different sensor information, different signals and also different algorithms and functions. These algorithms and functions should be integrated in sufficiently independent hardware elements.

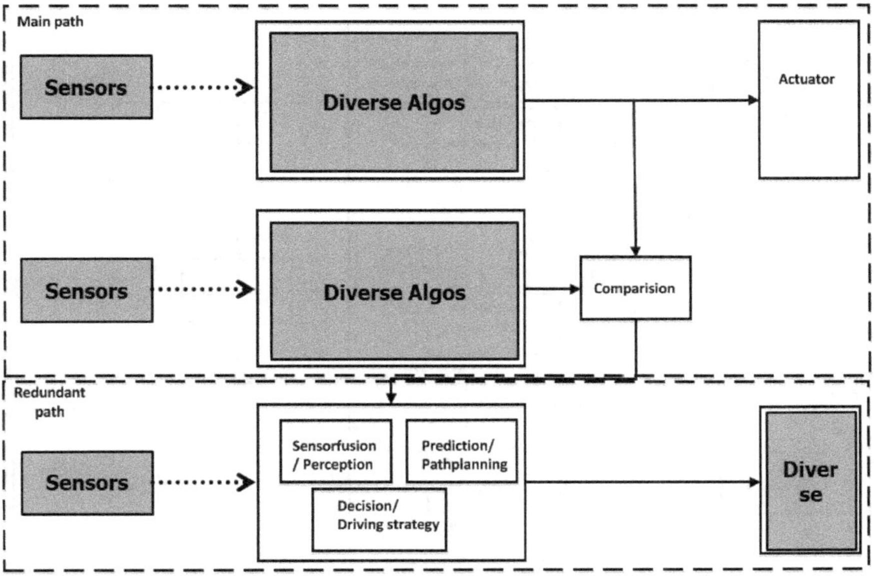

Figure 3.22: First approach: ASIL decomposition for ADAS/AD systems requires diversity

It is not always possible to use different redundant sensors because the sensor data from the camera, radar and lidar, for example, are generally collected in sensor fusion for the perception. There are also some other economical and also technological limitations which prevent having fully independent sensor data. All the sensor data from the camera, radar and lidar is usually used by the perception to detect static and dynamic objects. If the goal is to have fully independent decomposition paths as required by ISO 26262, each sensor must be integrated twice. But, if this is not applicable because of the cost, then the sensor independence shall be proved in a functional manner. Therefore, as shown in Fig. 3.23, it is recommended to use the same sensor data from the camera, radar and lidar for decomposition paths, but in different ways. The rule is that, if a failure occurs, the system remains safe on a specific decomposition path. For example, while the camera data is used as the main information for perception, the radar and lidar data is used as the main perception information for the other decomposition path.

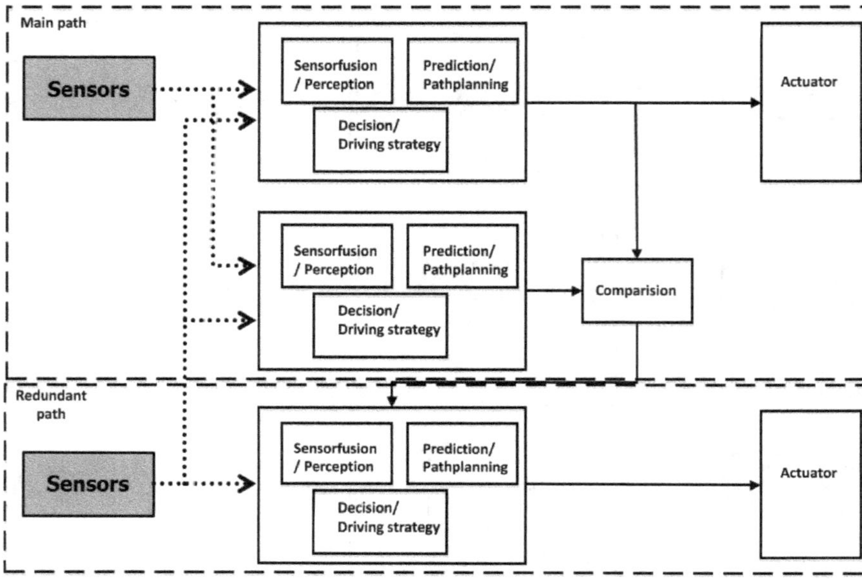

Figure 3.23: First approach: ASIL decomposition for ADAS/AD systems with reduced sensors

If the same sensor data is used for different decomposition paths, then it is very important, and also required by ISO 26262, to prove the sufficient independence of these paths as shown in Fig. 3.24. The dependent failure analysis plays a significant role in this proof. The DFA should be performed systematically and should also be performed in a different way than it is usually done today. The details for the DFA are explained in subsection 3.4.3.

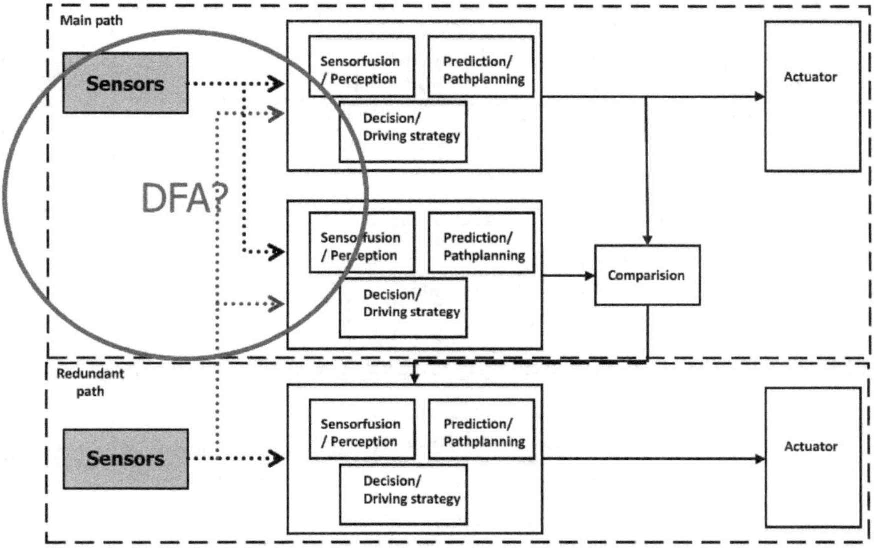

Figure 3.24: First approach: The necessity of DFA for ADAS/AD systems decomposition

In contrast to the approach illustrated in Fig. 3.22, the signal output/vehicle behavior of the actuators is checked for correlation with the input signals from the sensors within the second decomposition path. If the check results in implausibility, the second decomposition path switches to the redundant path.

As shown in Fig. 3.25, it is necessary to develop comprehensive monitoring functions for this architecture concept, to cover all aspects of the main functions and to detect any possible implausibility that could lead to a safety goal violation. These monitoring functions must be integrated in the sufficiently independent hardware elements. It is also necessary to perform a detailed DFA to prove the independence of the decomposition paths, analog to the first approach. The details for the DFA are explained in chapter 3.4.3.

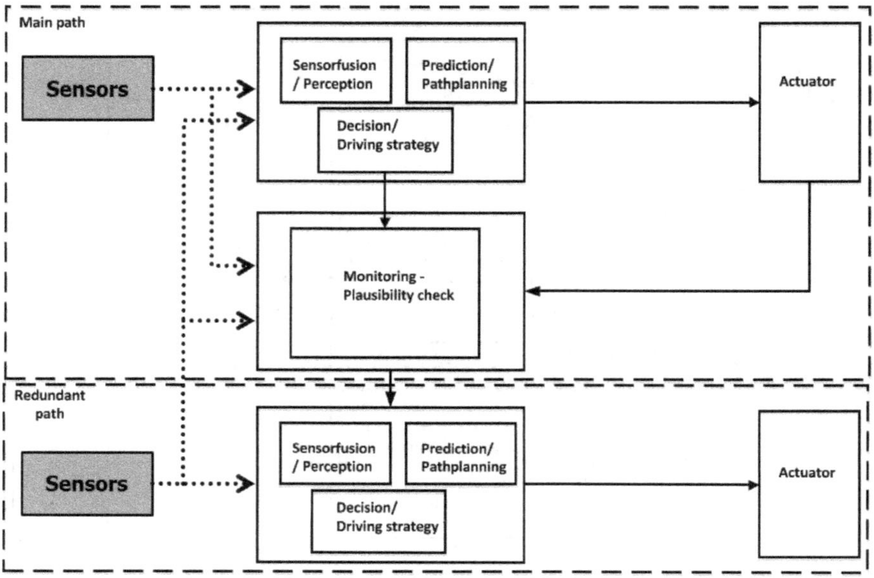

Figure 3.25: Second approach: ASIL decomposition for ADAS/AD systems

3.4.3 Dependent Failure Analysis for ADAS/AD Systems

The DFA for ADAS/AD systems entails additional challenges because of the SOTIF-related performance limitations.

The dependent failure analysis is used to document the freedom from interference, absence of common cause failures and the absence of SOTIF-related technological shortcomings, which are needed to prove independence when ASIL decomposition is being used. Fig. 3.26 illustrates the extended DFA methods with the SOTIF-related technological limitations. While common cause failures and cascading failures are already investigated for independence, the SOTIF-related triggering events should be also evaluated regarding the independence of the decomposition paths.

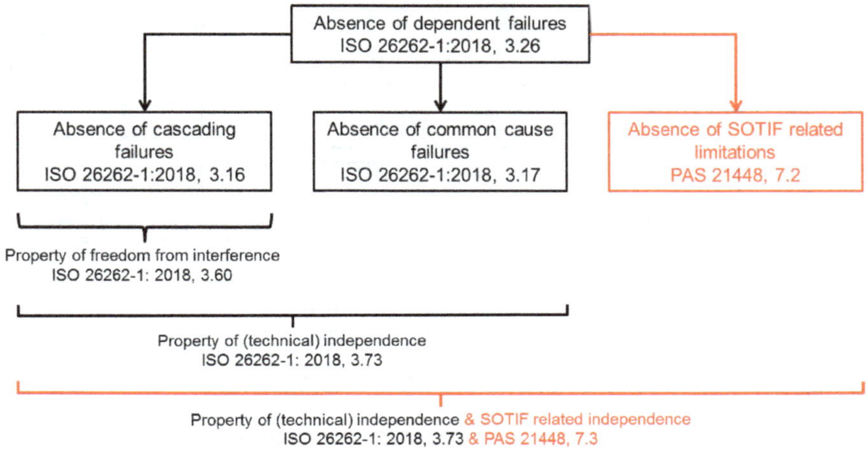

Figure 3.26: Extended DFA

The Fig. 3.27 shows the additional SOTIF-related dependent failure initiators (DFI). There are two additional DFIs, SOTIF-related triggering events involving the sensors, weather conditions for example, and SOTIF-related triggering events involving the algorithms for the environment and location. The DFA is performed according to the SOTIF-related triggering events in order to evaluate the dependencies between the decomposed functions and also for the redundant path.

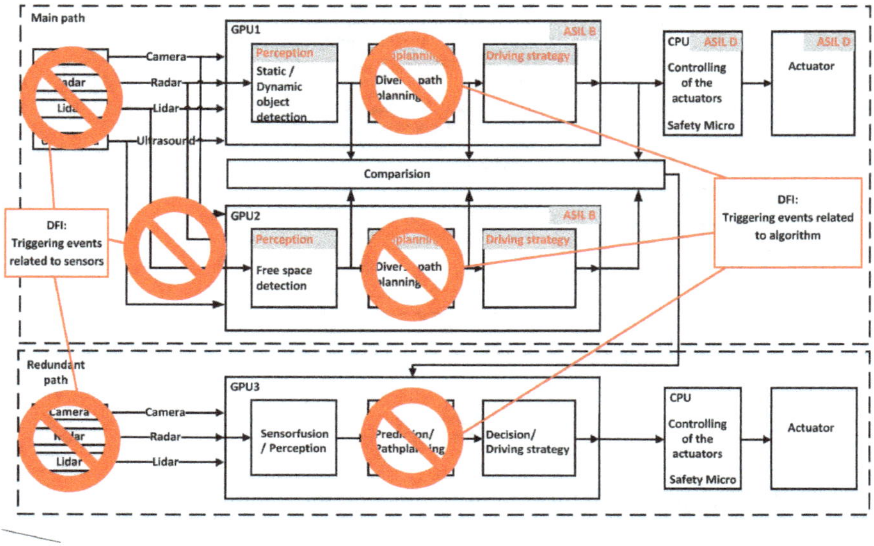

Figure 3.27: Dependent failure initiators for ADAS/AD systems

Chapter 4 describes a systematic development strategy for a model-based and automatically-performed dependent failure analysis. The use of DFA for SOTIF-related dependent failure initiators is also shown in subsection 4.2.3.

3.5 Use Cases

3.5.1 Fail-operational Safety Architecture for Powertrain Domain

The fail-operational safety architecture concept developed for domain ECUs can be applied, as shown in Fig. 3.28 to a power train domain. The transmission control ECU, electric machine and inverter ECU can be integrated to form a powertrain domain ECU, which then consists of two independent and diverse multicore processors. It is important to use two different multicore processors to reduce systematic errors from the semiconductor supplier. Each processor has its own redundant power supply and redundant signals for the execution of the software. Each processor is also monitored by a different independent watchdog. If a failure occurs on one of the processors, then the safety critical functions of this processor are performed by the backup core of the other processor. In this way, the system remains fail-operational as long as the backup core is able to execute the software correctly.

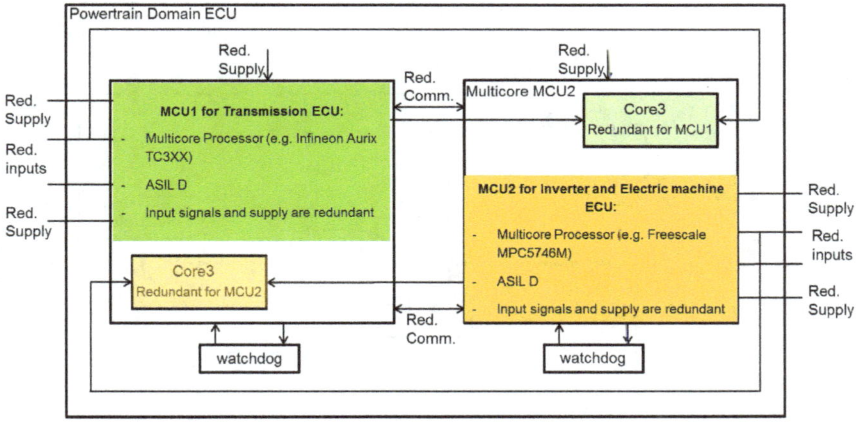

Figure 3.28: Fail-operational safety architecture for powertrain domain

3.5.2 ASIL Decomposition in General

The decomposition example from ISO 26262–part10, clause11.3 is continued in Fig. 3.29 to document the suitability of the fail-operational safety architecture within the domain ECU for the application of the decomposition approach. Fig. 3.29 shows the example extended for a system with an actuator which is triggered by driver demand to open the door using a dashboard switch.

For the purpose of this example, the architecture of the item is extended as follows:

- The driver request is read by the actuator control ECU, which powers the actuator through a dedicated power line.

- The vehicle speed information is provided by the VS ECU via CAN communication and is compliant with ASIL A.
- The second vehicle speed information is provided directly by the speed sensor and is compliant with ASIL B.

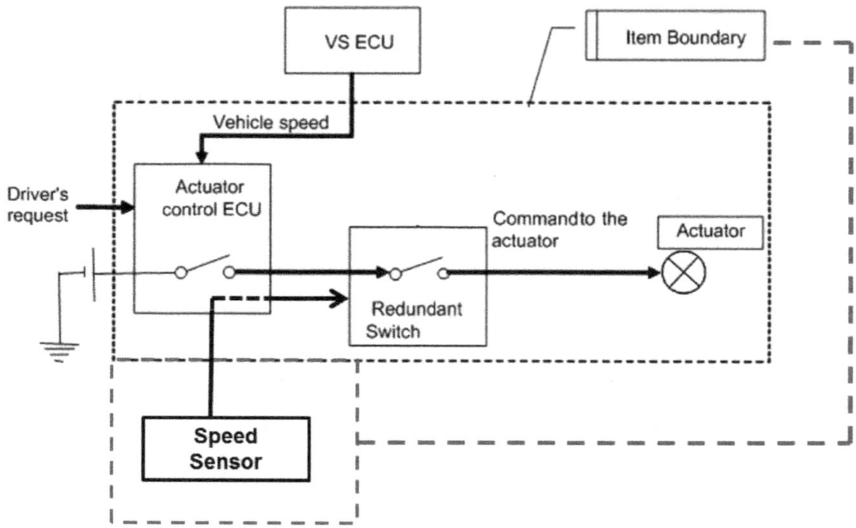

Figure 3.29: Extended decomposition example from ISO 26262-part10

The **hazardous event** considered in the analysis is the opening of the vehicle door caused by the activation of the actuator while driving at a vehicle speed above 15 km/h, with or without a driver request. For the purpose of the example, the ASIL rating associated with this hazardous event is ASIL C. The corresponding **safety goal** is "Prevent the door from opening due to actuator activation at a vehicle speed greater than 15 km/h".

The defined safety goal can be fulfilled by the implementation of the following decomposed requirements:

- Requirement 1: the actuator control ECU shall not power the actuator if the **vehicle speed from the VS ECU** is greater than 15 km/h => ASIL A(C).

- Requirement 2: The redundant switch shall be in an open state if the **vehicle speed from the speed sensor** is greater than 15 km/h. => ASIL B(C).

As described in subsection 3.4.2.1, the safety goal "Prevent the door from opening due to actuator activation at a vehicle speed greater than 15 km/h" can be realized through the application of ASIL decomposition. The fail-operational safety architecture illustrated in this example enables the decomposed safety requirements allocated to architectural elements to be sufficiently independent.

This example is examined further for model-based DFA in subsection 4.2.6.

3.5.3 ASIL Decomposition for AD Systems

Most of the malfunctions related to highly autonomous driving are classified with ASIL D, depending on the scenario. In contradistinction to the decomposition approach for existing systems, there are additional challenges for the application of the decomposition approach for ADAS/AD systems. The solutions/processing steps for using decomposition are explained below in the context of the example in Fig. 3.30:

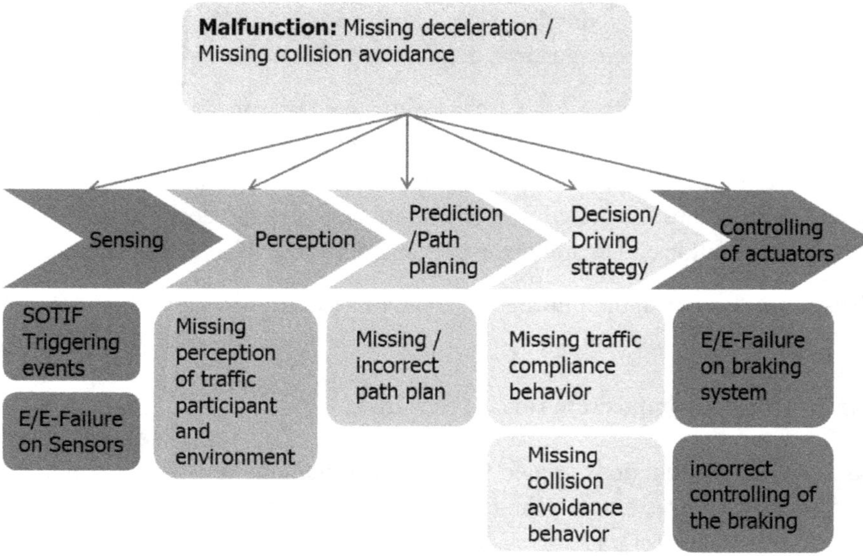

Figure 3.30: Use case: Failure causes for missing collision avoidance

Function: Collision avoidance system
Malfunction: Missing deceleration/Missing collision avoidance behavior. The system fails to activate collision avoidance functions when required. For example, the vehicle does not stop for a pedestrian. This malfunction could cause life-threatening injuries or fatal injuries and fewer than 90% of all drivers or other traffic participants are usually able to avoid harm or control the vehicle. So this kind of scenario is evaluated with an ASIL D rating (Exposure: 4, Severity: 3, Controllability: 3).

Causes for this malfunction in the AD processing chain:

- E/E failure in the braking system. The system fails to activate the braking system.
- Incorrect traffic compliance behavior. The system fails to drive in accordance with traffic compliance behavior. For example, the vehicle fails to stop at a traffic light.

- Missing collision avoidance function. The system fails to activate collision avoidance behavior. For example, the vehicle fails to stop for a traffic participant.
- Incorrect path plan calculation. The system fails to calculate/predict a correct path plan due to the wrong localization of the vehicle.
- Missing/incorrect perception of traffic participant and environment. The system fails to recognize the traffic participants correctly.
- E/E failure in the sensing system. The sensors fail to collect the correct environmental data to recognize traffic participants.
- Performance limitation / technological shortcomings (SOTIF triggering events) on the sensors. The sensors are not able to provide the correct data to the perception algorithms because of the sensor limitations.

Safety goal: Collision avoidance shall be ensured (ASIL D). In order to achieve this safety goal, the following top level functional safety requirement is defined:

Top level safety requirement: The autonomous driving system shall ensure the correct control of the actuators, based on the input, to avoid collisions with traffic participants, static and dynamic obstacles (ASIL D).

To fulfill the corresponding safety requirements, the algorithms and related hardware must also be developed in ASIL D, as shown in the following Fig. 3.31. But the hardware from the sensors to the high performance chips, where the AD algorithms are implemented, is not adequate enough to achieve the ASIL D rating. This is why decomposition is necessary to implement ASIL D safety goals for AD systems.

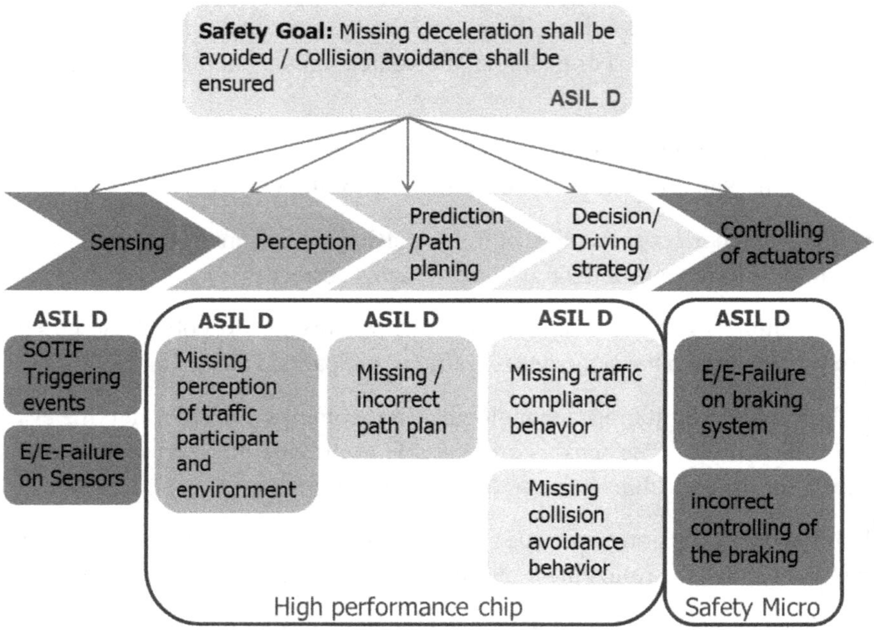

Figure 3.31: Use case: ASIL rate for related functions

Decomposition for AD systems presents some additional challenges in comparison to its use for existing systems. The decomposition concepts which were developed for AD systems are illustrated in the following Fig. 3.32.

In the first approach, the sensor signals are used for both decomposition paths, but in a way that the weight of the sensor signals is divided between the paths in consideration of the common cause failures. This means that an electrical failure or a SOTIF-related performance limitation of the sensors should not lead to a safety goal violation due to the simultaneous failure of both decomposition paths. The perception algorithms should be developed diversely in order to achieve different weightings of the sensors. For example, while the first decomposition path could detect the static and dynamic objects, the second decomposition path could calculate the free space detection. The results are compared to see whether they correspond. The path planning and driving strategy are also developed diversely. If the results are not plausible with each other, the system switches to the redundant path to remain fail-operational.

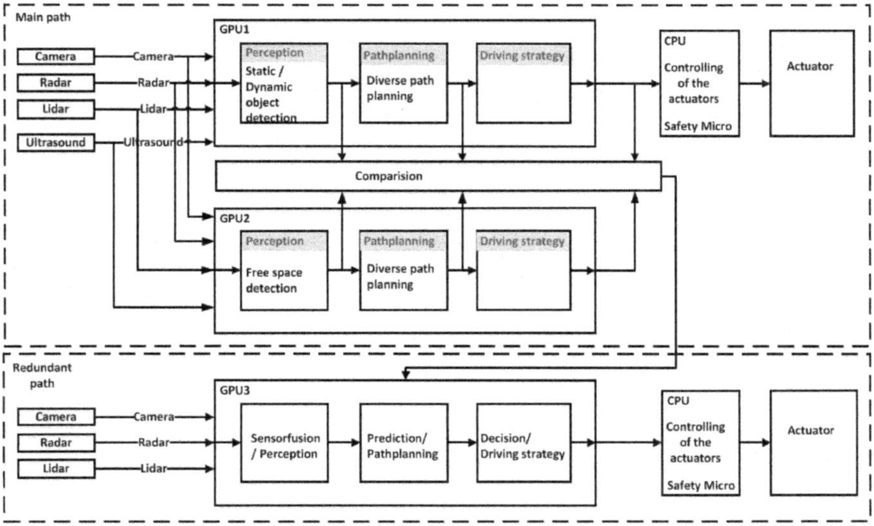

Figure 3.32: Use case: First approach for perception decomposition

As compared to the first approach, in the second approach the second decomposition path is used to check the plausibility of the system reaction to the input signals as shown in Fig. 3.33. For this reason, the actuator output signals are provided to the second decomposition path. The special functions in the second decomposition path check the compatibility of the first decomposition path results with the actuator output, based on the input signals which are used to calculate the expected system reactions. If the results are not compatible with each other (not plausible), then the system switches to the redundant path in order to remain fail-operational.

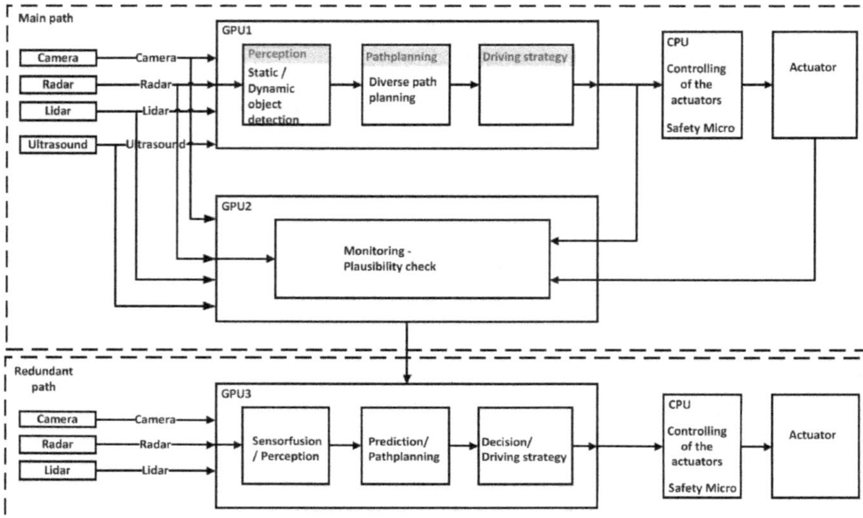

Figure 3.33: Use case: Second approach for perception decomposition

3.6 Conclusion

This chapter lays out the fail-operational safety architectures which were developed in consideration of domain ECUs within multicore processors. The methods resulting from this research provide a redundant system architecture and safety architecture for ADAS/AD systems with regard to their processing chain. The ADAS/AD processing chain consists of sensors, domain ECUs and the control of the actuators. This research shows how these elements can be designed in a fail-operational manner and how the redundancies of the design elements can be identified. It is important to use differing high performance chips and multicore processors from different suppliers to reduce the systematic failures. The proposed approach also enables the application of ASIL D safety requirements and increases the system availability with fail-operational functionality. The main advantage of the concept is the fail-operational realization of the electronic systems of the vehicles from sensor to actuator. This chapter also provides various solutions for using the decomposition approach and also for the extension of DFA considering SOTIF-related technological insufficiencies for safety critical ADAS/AD systems. The decomposition and extended DFA approaches are described in detail with the corresponding use cases. Finally, this chapter shows how a suitable fallback strategy is designed to achieve the minimum risk condition and to increase the availability in case of a system failure or system limitation.

4 Model-driven Approaches for ISO 26262 Work Products and DFA

This chapter presents two innovative approaches. The first approach describes the model-based development of ISO 26262 work products. The second approach concerns the model-based DFA which is required by ISO 26262 for the application of ASIL decomposition.

The application of the approaches is independent of the tools being used. The EAST-ADL standard is used for the evaluation of the approaches, so EAST-ADL modeling elements are developed further to enable the elaboration of these model-based approaches and to achieve the benefits of the solutions.

4.1 Development of Safety Functions Using Modified EAST-ADL

This subchapter presents a method for model-based development of system, software and safety architecture using the extended EAST-ADL standard in consideration of automotive safety standard ISO 26262. In particular, this subchapter includes a brief discussion of how the main safety work products required by ISO 26262, for example, the hazard analysis and risk assessment, the functional safety concept, the technical safety concept and the safety analysis, can be developed model-based, and how these safety activities can be combined with the system architectural design and linked to each other.

4.1.1 Introduction

This research document presents an approach for model-based development of safety critical functions and model-based development of ISO work products. EAST-ADL is used as the backbone for this approach, but EAST-ADL is mod-

B. Sari, *Fail-operational Safety Architecture for ADAS/AD Systems and a Model-driven Approach for Dependent Failure Analysis*, Wissenschaftliche Reihe Fahrzeugtechnik Universität Stuttgart, https://doi.org/10.1007/978-3-658-29422-9_4

ified in a way which brings the system architecture and safety architecture together, because ISO 26262 requires consistency and traceability during the development of safety critical systems. The current, unmodified EAST-ADL specification is not adequate for the implementation of these requirements, because the system architecture and safety architecture are separated into different models and there is no direct relationship between the system architecture models and the safety models. The system architecture is developed within the abstraction levels of EAST-ADL and the safety models are realized within the dependability model. It is a big challenge, and also a requirement of ISO 26262-part4, Figure2 and ISO 26262-part10, Figure8 and Figure9, to show the dependencies between HARA, safety goals, functional and technical safety requirements and safety functions. Using safety assessments, the developers should be able to show which safety functions implement which safety goals and be able to prove that the safety functions and safety goals have the same ASIL. It was, therefore, essential to modify EAST-ADL in a way which brings the system and safety models together to provide the required consistency check, the traceability of safety functions and also the automatic generation of the ISO 26262 work products. The extensions to EAST-ADL enable the representation of the relationship between the safety goals and the safety functions, so it is easy to check which function is implemented for which safety goal. The modification of EAST-ADL makes it easier to prove the completeness of the safety activities.

The second main contribution of this approach is to verify and to validate the technical safety concept and functional safety concept, as required by ISO 26262-part4, clause7.4.8, using the simulation of safety requirements in the earlier development phase of the project. Simulation of the safety requirements makes it possible to improve the safety concepts in an earlier phase of development and enables the earlier detection of systematic errors. The modeling of a complete system is an extensive project and requires domain specific know-how. Knowledge of various tools is also needed, because the architecture is described using several different tools. The current approach can be replaced by the approach developed in this research, which is based on a domain-specific language with ADL [21]. The proposed approach, with its extensions and improvements, makes it possible to create a system model containing all the pertinent information and which describes the complete system, safety and software architecture and offers the required consistency and traceability between system and safety architecture and enables the verification of the safety concepts in an earlier development phase.

The following subchapters show how EAST-ADL is modified for this purpose.

4.1.2 Description of the Approach

The developed approach in Fig. 4.1 shows how the system, hardware, software, and safety architecture development are combined, which decreases the complexity of the system by providing a coherent architecture.

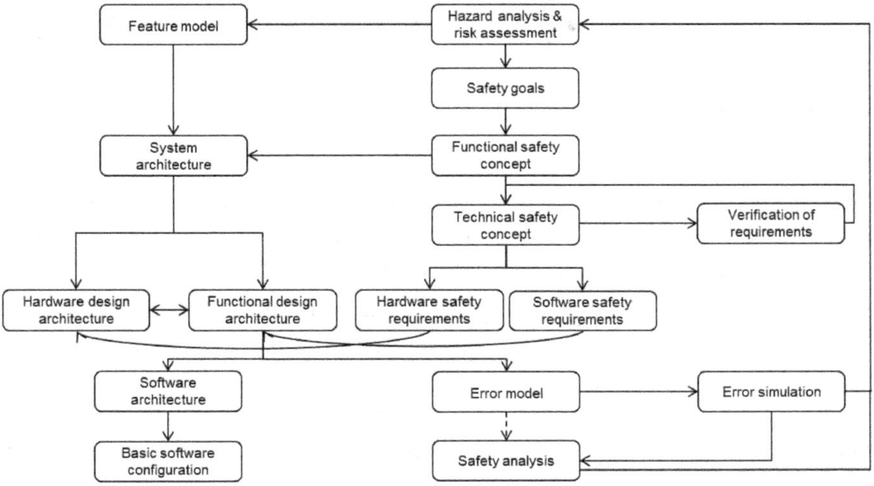

Figure 4.1: Model driven approach for system, hardware, software and safety architecture development

The developed method consists of five main parts, which are shown in Fig. 4.2.

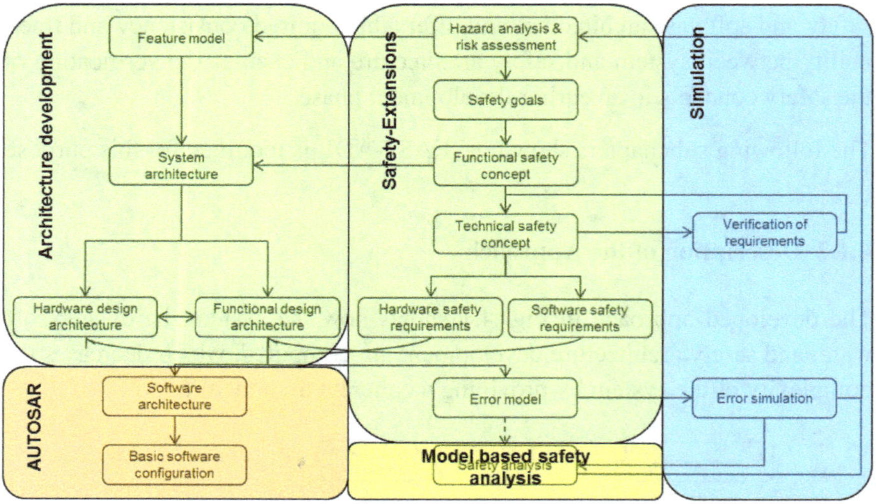

Figure 4.2: Main parts of the approach

The first part of the approach is the **architecture development** [4]. This part concerns the creation of the feature model, system architecture, FDA and HDA. It is also possible to allocate the functions from the FDA model to the corresponding hardware elements of the HDA model. The architecture development is widened to include additional safety attributes in order to combine the system architecture and safety architecture. This allows representation of the relationship between the system architecture and the safety architecture, which enables the proof of traceability and consistency. The further details are explained in subsection 4.1.3.1.

The second part of the method is **safety extensions** [2, 35]. This part deals with the model-based creation of ISO 26262 work products as an extension to architecture development. Firstly, hazard analysis and risk assessment are performed. Secondly, the safety goals are derived from the HARA. In the following step, the functional safety concept and the technical safety concept are created to fulfill the safety goals. This part is also enhanced to include additional safety attributes to form a relationship between the safety model, the system model and the requirements model. The safety model enables the model-based realization of safety work products. Now additionally, using the developed scripts, the safety extensions enable these safety work products to be automatically generated from the models. Further details are explained in subsection 4.1.3.2.

The feature model can contain both non-safety-critical and safety-critical properties of the system. After the HARA, the safety-critical aspects of the system are considered as being part of the feature model. In the system architecture, the safety goals and corresponding safety functions of the system are taken into account. The safety-critical functions are detailed in the functional and hardware design architecture.

The third part of the approach is **AUTOSAR** [16]. This part concerns the software architecture and basic software configuration. The software architecture can be created from the FDA model, which contains the necessary software features.

The fourth part is the **model-based safety analysis**. In this step, the FTA of the fault models are automatically generated, using external tools, from the error model of the ADL or from simulation model [5]. The further details are explained in subsection 4.1.3.4.

The last part of the method is **simulation and verification**. This part is developed within this research. Error simulation and verification of the requirements are carried out in this step, and these tasks are made possible in an earlier phase of development. On the one hand, the safety requirements are verified by using a simulation environment. On the other hand, it is possible to determine the system impacts of causes with error simulation. Thus, the safety goals of the system, which were defined by the HARA, can be approved. Further details are explained in subsection 4.1.3.5.

The developed approach allows for the consistency and traceability of the individual steps. The method also permits efficient tracing from the software architecture to the feature model and also from the safety analysis to the HARA.

4.1.3 Extensions of EAST-ADL

4.1.3.1 Extensions of EAST-ADL Abstraction Level:

The library elements of these abstraction levels are extended as shown in Fig. 4.3. The EAST-ADL abstraction layer elements contain additional information about the corresponding safety goals, safety requirements and ASIL classification, which enable the representation of the relationship between the safety model and the system model. Additionally, the attribute "verified" shows whether the function is already verified in an earlier development phase. This information is very useful for safety case generation and is also required by ISO 26262-part4, clause7.4.8.1.

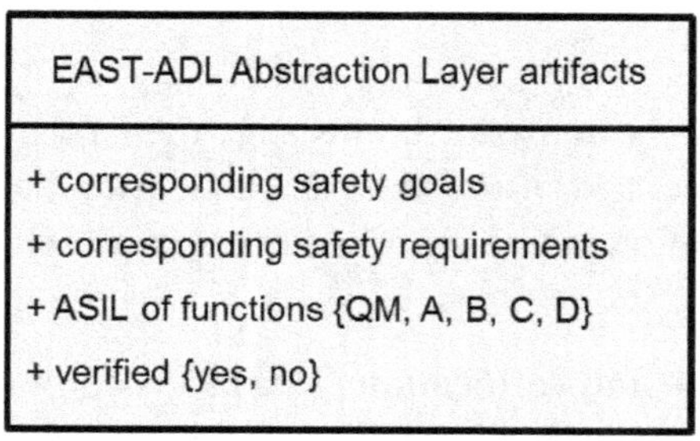

Figure 4.3: The extensions of EAST-ADL abstraction layer

4.1.3.2 Extensions of EAST-ADL Dependability Model:

EAST-ADL offers the dependability package with safety extensions that enable the work products of the safety development process to be created, such as the HARA. The dependability model is additionally designed to support the developer in creating safety requirements and performing the safety analysis, as well as in the modeling of systematic errors and their propagation, which leads to failures. The library elements of the dependability model were also widened in this research, as follows in Fig. 4.4.

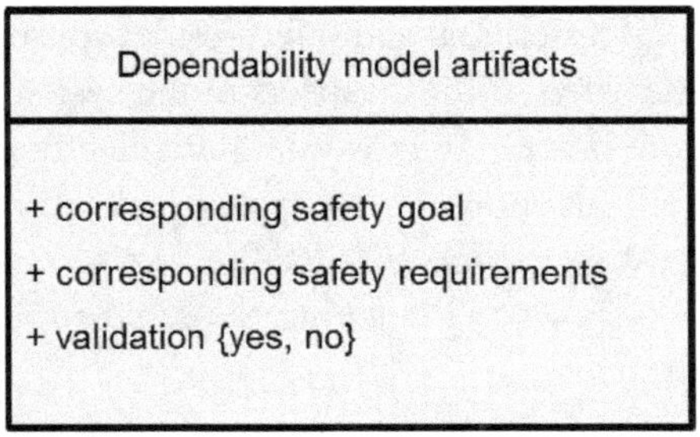

Figure 4.4: The extensions of EAST-ADL dependability model

The extensions added in this research enable the generation of safety concepts using the automation scripts which were also developed. The safety concepts are created automatically from the modified requirements model and the dependability model shown in Fig. 4.5.

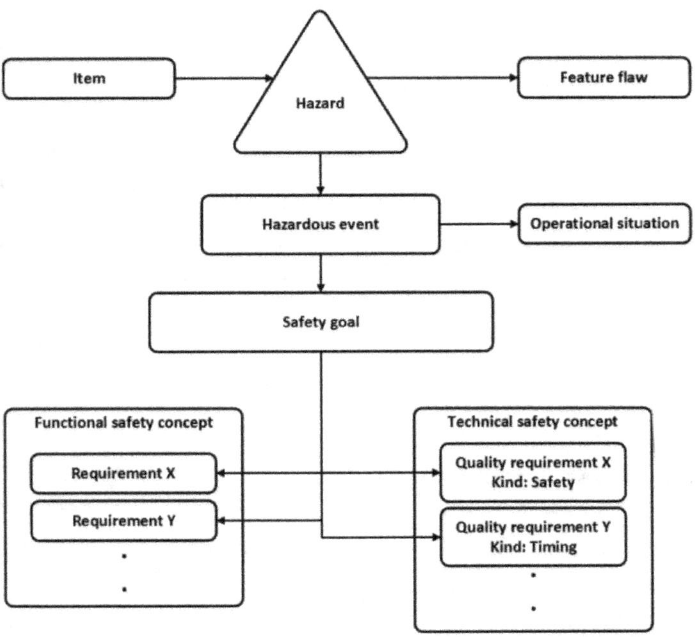

Figure 4.5: EAST-ADL dependability model

The feature model can contain both non-safety-critical and safety-critical properties of the system. After the HARA, the safety-critical aspects of the system are considered as being part of the feature model. In the system architecture, the safety goals and corresponding safety functions of the system are taken into account. In the functional and hardware design architecture, the safety functions continue to be refined. Fig. 4.6 shows the mapping of architecture development artifacts and safety work products.

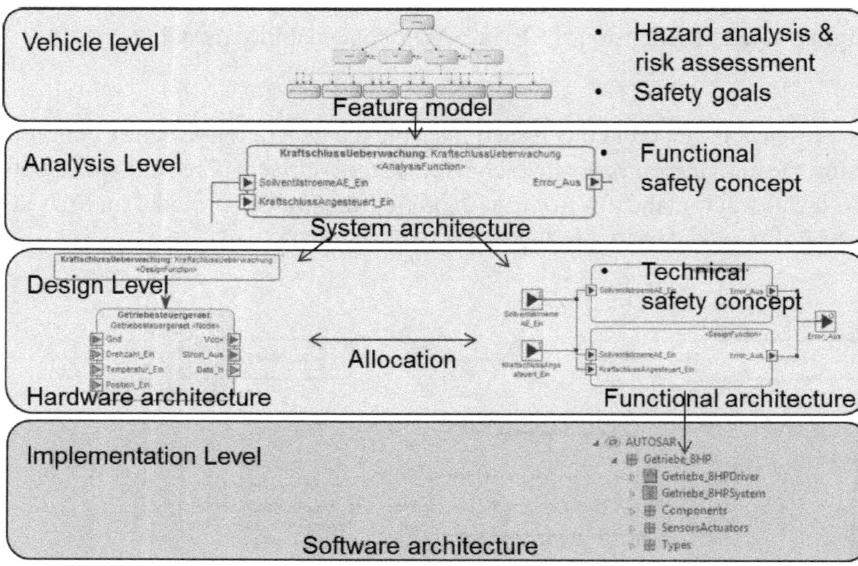

Figure 4.6: Mapping of the EAST-ADL levels and ISO 26262 safety work products

4.1.3.3 Extensions of EAST-ADL Requirements Model:

During the development of a system, requirements should be defined that can be modeled in a separate requirement diagram as shown in Fig. 4.7. The requirements model can include both non-safety-critical and safety-critical requirements, so it is recommended to begin the requirements modeling on the system architecture level.

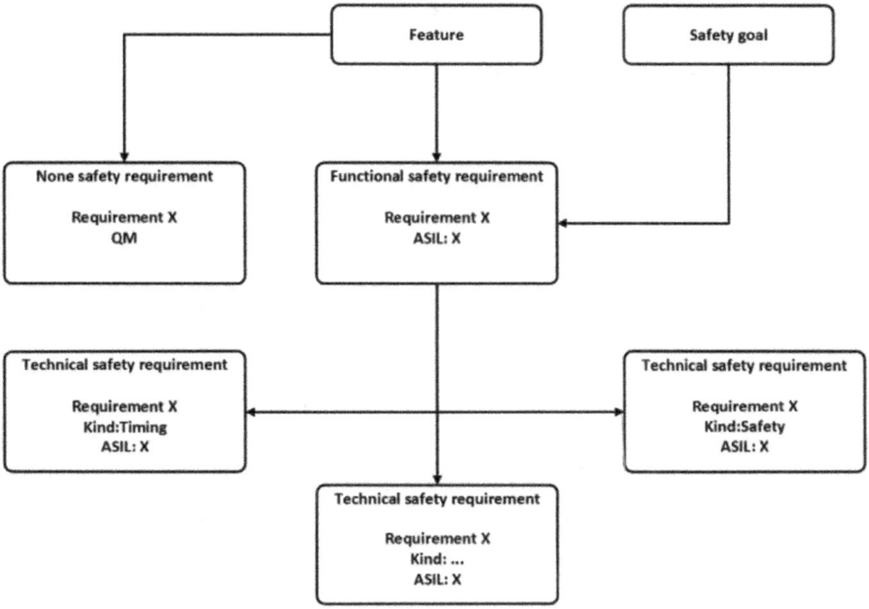

Figure 4.7: EAST-ADL requirements model

The requirements model in EAST-ADL is based on the SysML, which is adapted to the metamodel of EAST-ADL. User-defined features can be added to the requirements, as it is also possible in DOORS. Information is added, e.g., status, author and responsible person, to achieve better traceability of the changes. The requirements model can be exported in the ReqIF format to facilitate exchanges with other requirements tools. As mentioned before, the requirements model is enhanced, as shown in Fig. 4.8 with safety-related features such as the ASIL classification, the corresponding safety goal or the decomposition relevance.

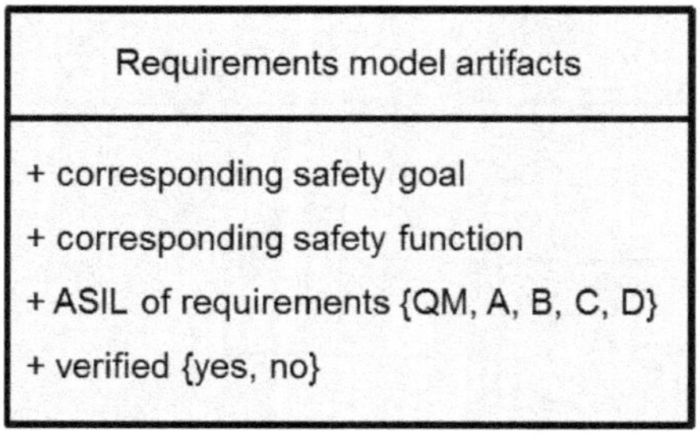

Figure 4.8: The extensions of EAST-ADL requirements model

The extensions, which were added in this research, enable the automatic generation of safety concepts using the developed script, as follows in Fig. 4.9.

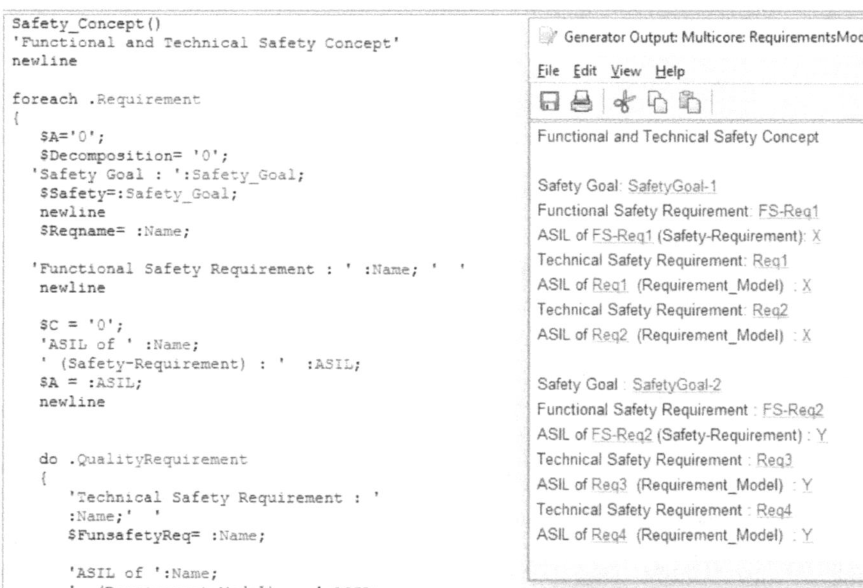

Figure 4.9: Safety concept script and generated document

It is possible to automatically generate error models from the FDA. The error description and error logic for the individual subsystems are modeled afterwards. Vehicles in general, and especially electrical vehicles, consist of complex systems, so it is very difficult to determine all the measures which are necessary to prevent a hazard. Of course, OEMs and their system suppliers are committed to reduce all risks to an acceptable level. This means that, for all hazards and risks, the relevant safety goals, safety concept, safety requirements and safety measures must be defined. For this reason, ISO 26262 requires the safety analysis to be carried out based on the fault models in the automotive industry. For example, with the help of the FTA, it is possible to find the error causes for a particular hazard [14]. In EAST-ADL, it is possible to model errors and their propagation in an error model to recreate a failure behavior. On the basis of error models, additional tools such as HiP-HOPS [5] can be used to perform model-based safety analysis. Fig. 4.10 can be used to perform model-based safety analysis.

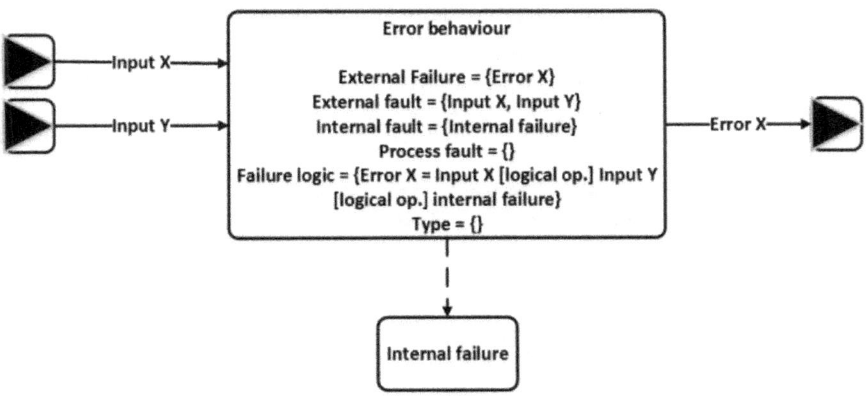

Figure 4.10: Error model

4.1.3.4 Model Based Safety Analysis:

It is possible to use commercial tools such as HiP-HOPS [5] to perform the safety analysis. These tools are capable of automatically generating the FTA and the FMEA from an error model.

The safety analysis enables discovery of the possible error causes which should be detected and prevented with the appropriate safety measures. For this purpose, the safety measures and requirements are defined in order to introduce countermeasures. Additionally, functional and technical safety requirements can be defined or extended, based on the results of the safety analysis.

4.1.3.5 Simulation

For the verification of the requirements and error simulation, a simulation environment such as Simulink (Matlab), Dymola, CarMaker, etc. can be used. For instance, the powertrain of an electric vehicle can be simulated with the help of a simulation environment and error models can be designed and implemented. The behavior of the vehicle can be simulated with error models in such a way that the causes for the error, e.g. failure sensor values, can be used as a stimulation signal. The simulation is used for the verification of the requirements, for the detection of errors in an earlier phase of development and for detection of the impacts errors have on the system. If it is found that the safety requirement cannot be implemented as intended, then the requirement should be refined. This is shown in Fig. 4.11.

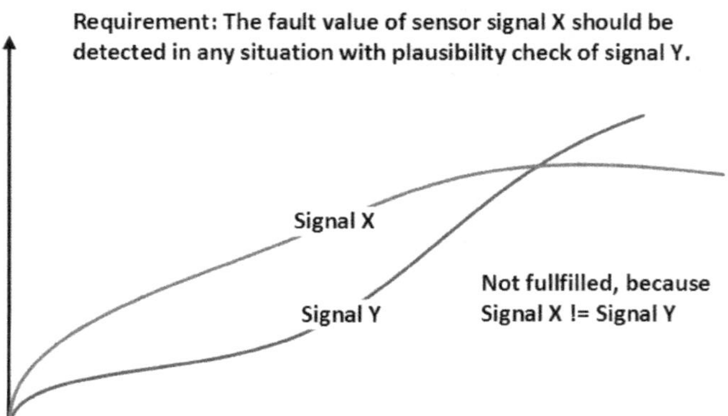

Figure 4.11: Verification and simulation

Fig. 4.12 shows the evaluation procedure for the defined hazards. The simulation is performed using the error causes to find out the system reaction to these failures. If the vehicle behavior is the same as in the defined hazards, then the HARA is approved, otherwise the HARA should be extended further as required by ISO 26262-part4, clause7.4.8 to develop the necessary safety measures for detecting and preventing the possible hazards.

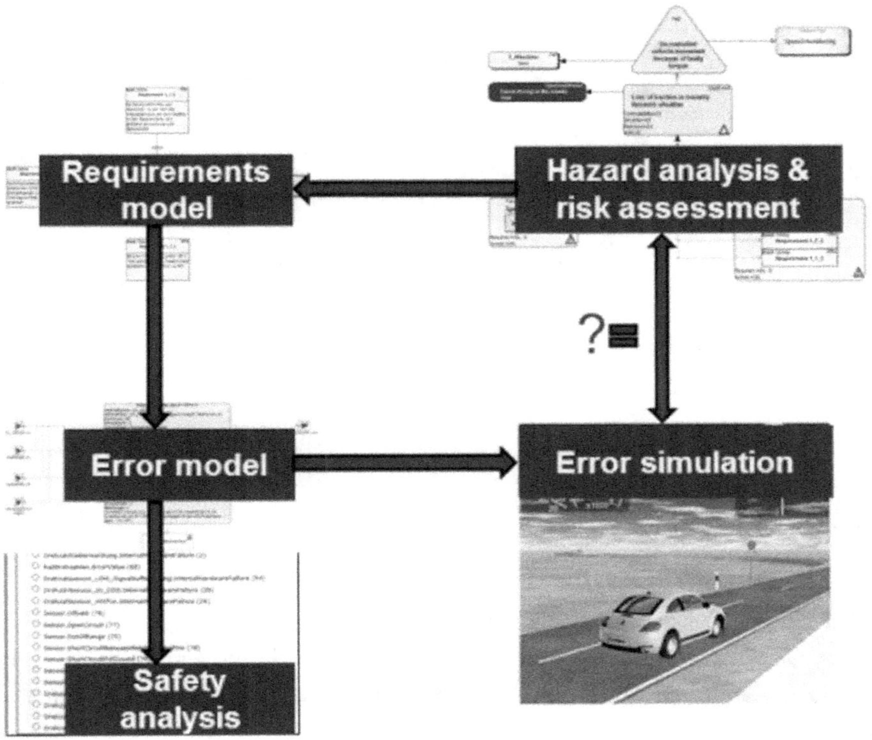

Figure 4.12: Error simulation

4.1.4 Use Case

As an illustration of the benefits of the developed model-driven approach, an example is used which shows the safety-critical function development. The example starts with the HARA and ends with the verification of the safety goals and the safety requirements. Fig. 4.13 shows a typical dependability model

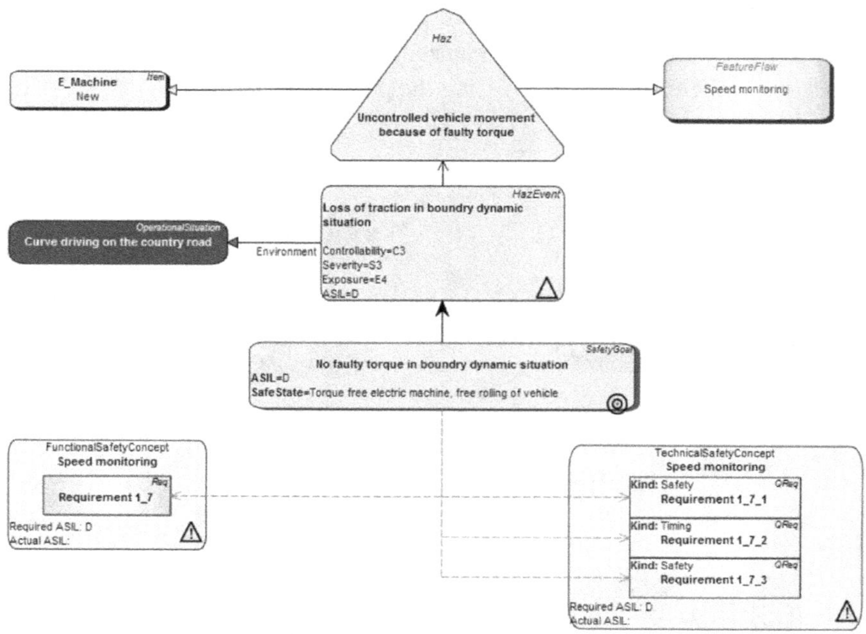

Figure 4.13: Dependability model – power train example

of a system. In this concrete case, an electric machine is considered as an item. The hazard is defined as "Uncontrolled vehicle movement because of faulty torque". The ASIL of the hazardous event is classified based on the operational situation. In this case, the hazardous event "Loss of traction in boundary dynamic situation" leads, during driving in a curve, to a dangerous situation. Therefore, the ASIL of this hazardous event is classified as ASIL D (C3, S3, E4). In the next step, the safety goal "No faulty torque in boundary dynamic situation" is defined with the safe state "Electric machine is torque-free, vehicle is rolling freely" in order to detect and prevent the failure. The functional safety concept and the technical safety concept are then created to achieve the safety goal. The functional and technical safety requirements are linked from the requirements model, and should not be re-specified in the dependability model.

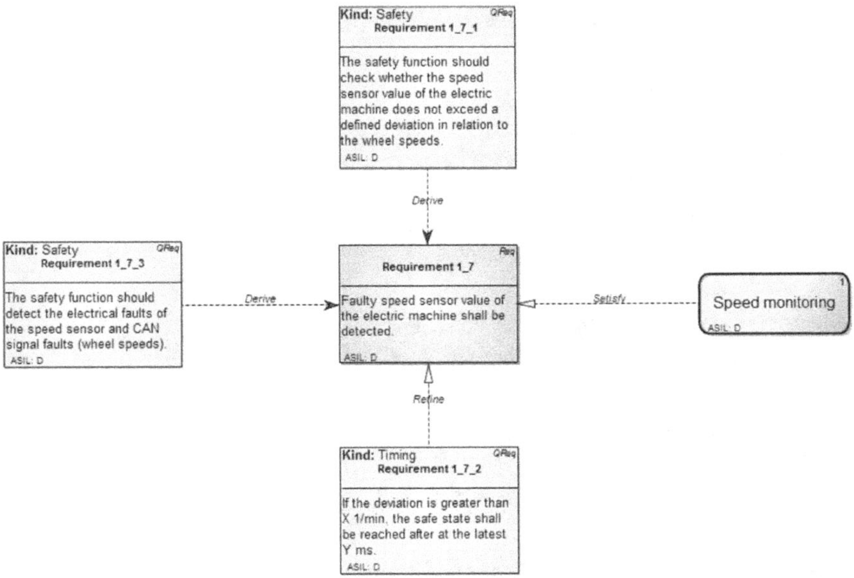

Figure 4.14: Requirements model – power train example

Functional and technical safety requirements are specified using the associated information from within the requirements model. The safety requirements are allocated to the architectural elements by using the "satisfy" relationship. In this example, the safety goal is implemented with the function "Speed monitoring", as shown in Fig. 4.14.

The functional and technical safety concept can be automatically created from the dependability model and requirements model using the developed script, as shown in Fig. 4.15.

Fig. 4.16 shows the FDA model. The safety requirements are implemented within the subfunctions of the FDA model, which are enhanced with safety attributes to enable consistency and traceability check.

The error model of the hazard is generated from the FDA-Model automatically. Afterwards, the error logic of the subsystems is described individually with the possible fault causes and considering error propagation. The error logic

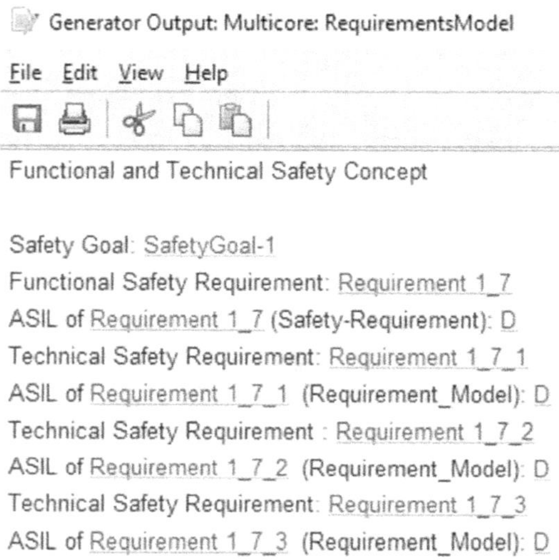

Generator Output: Multicore: RequirementsModel

File Edit View Help

Functional and Technical Safety Concept

Safety Goal: SafetyGoal-1
Functional Safety Requirement: Requirement 1_7
ASIL of Requirement 1_7 (Safety-Requirement): D
Technical Safety Requirement: Requirement 1_7_1
ASIL of Requirement 1_7_1 (Requirement_Model): D
Technical Safety Requirement : Requirement 1_7_2
ASIL of Requirement 1_7_2 (Requirement_Model): D
Technical Safety Requirement: Requirement 1_7_3
ASIL of Requirement 1_7_3 (Requirement_Model): D

Figure 4.15: Functional and Technical Safety Concept – power train example

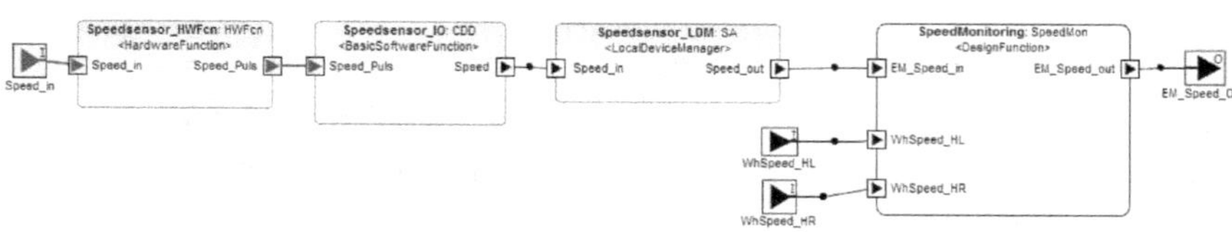

Figure 4.16: Functional design model – power train example

contains all the possible output failures of every submodule, with regard to internal hardware failures, internal software failures and input signal failures, as it is shown in Fig. 4.17.

The additional tool HiP-HOPS enables a model-based safety analysis to be performed automatically from the existing error models. This tool is capable of generating cut sets, FTA and FMEA. Fig. 4.18 shows the generated FTA for the top event “faulty torque”. The causes which can lead to the top event, e.g., sensor failures, internal software failures and internal hardware failures, are listed in the FTA. In this case, a qualitative safety analysis is performed. The

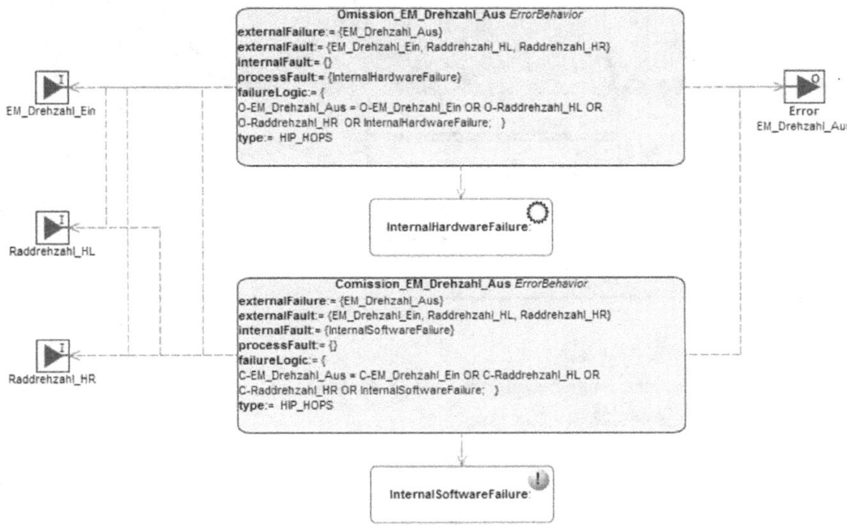

Figure 4.17: Error model logic – power train example

tool also supports the completion of a quantitative analysis.

Finally, the error simulation is performed, in order to verify the system effects (top events) which were previously defined in the HARA. For this reason, the simulation is stimulated using the error causes (basic events) from the fault tree analysis. The basic events should lead to the top events. After the simulation, the results are compared to the HARA results. If they are same, then the defined hazardous events are correct and thus verified. But if the vehicle behavior differs from the defined events, then the HARA should be extended. If necessary, a new safety goal should then be defined, in order to prevent the newly detected system impacts. In this example, a concrete hazardous event (faulty torque due to E-Machine speed sensor failures) leads to loss of traction, which causes the vehicle to enter an instable state. In the boundary dynamic situations, e.g. when driving in a curve, the driver can no longer control the vehicle. This can lead to a fatal accident. Fig. 4.19 shows both the verification results of the technical safety requirements and the validation of the safety goals with respect to the error simulation.

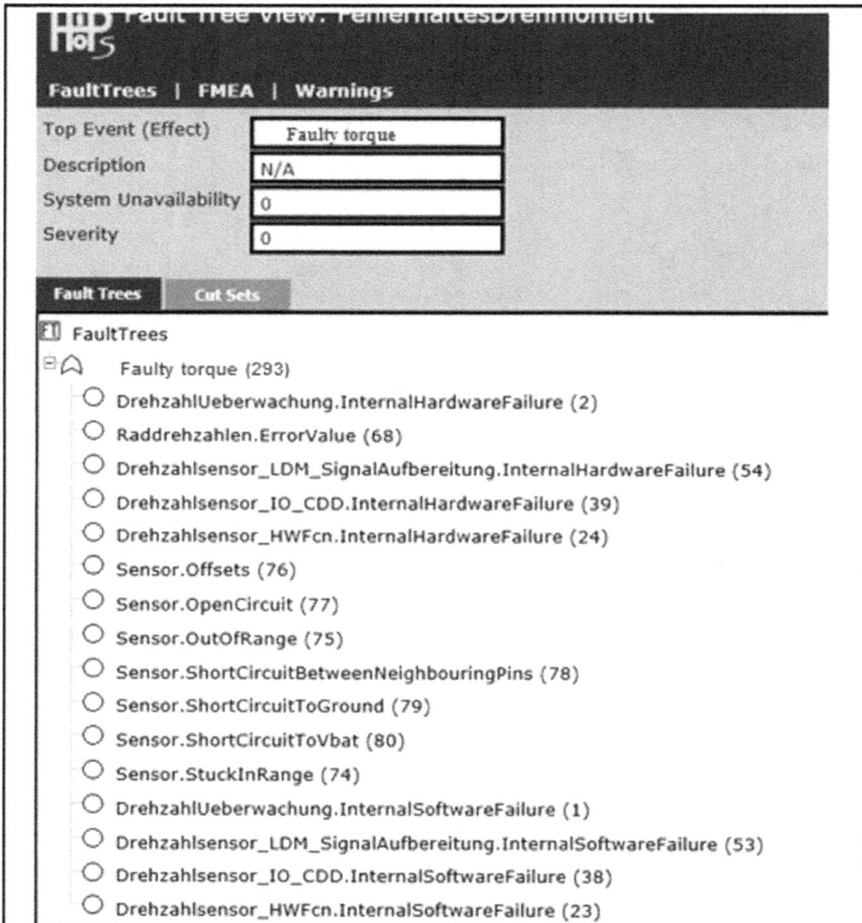

Figure 4.18: Model based safety analysis with HiP-HOPS – power train example

Verification and Validation

Safety Goal: SafetyGoal-1
Validated: yes

Functional Safety Requirement: Requirement 1_7
Verified: yes

Technical Safety Requirement: Requirement 1_7_1
Technical Safety Requirement: Requirement 1_7_2
Technical Safety Requirement: Requirement 1_7_3
Verified: yes

ASIL of FDA SpeedMonitoring : D
ASIL of FAA SpeedMonitoring : D
Verified: yes

ASIL of FDA Speedsensor HWFcn: D
ASIL of FAA FD: Speedsensor: D
Verified: yes

ASIL of FDA Speedsensor_IO: D
ASIL of FAA FD: Speedsensor : D
Verified: yes

ASIL of FDA Speedsensor_LDM: D
ASIL of FAA FD: Speedsensor: D

Figure 4.19: Verification and Validation – power train example

4.1.5 Conclusion

The approach researched in this dissertation supports developers in creating a model-based system, hardware, software and safety architecture for E/E systems. Various tools are currently used for different special development fields. The architecture description language EAST-ADL, the main component of this approach, replaces these tools in a way that the entire development of system and safety architecture can be realized within EAST-ADL. The proposed approach has the advantage of verifying the functional safety concept and technical safety concept and also of validating the HARA within the developed simulation environment in an earlier development phase of project.

The most significant advantage of the method presented here, is that it helps prove the traceability of the various development steps and safety work products with the aid of extended library elements and developed scripts. Once the user becomes familiar with the method, it is easy for him to understand the overall architecture. Fig. 4.20 highlights the relationship between the various levels and shows the traceability of the various steps.

The approach taken in this research enables the system, hardware, software and safety architecture of an item to be understood very quickly. When this approach is used, the requirement model, the dependability model and the library elements of EAST-ADL are extended with safety features. This enhancement enables the safety-critical requirements to be modeled with the necessary attributes, enables the functional and technical safety concepts to be created automatically using the developed script and enables the system and safety architectures to be combined. This permits the user to quickly identify which functions are implemented for which requirements.

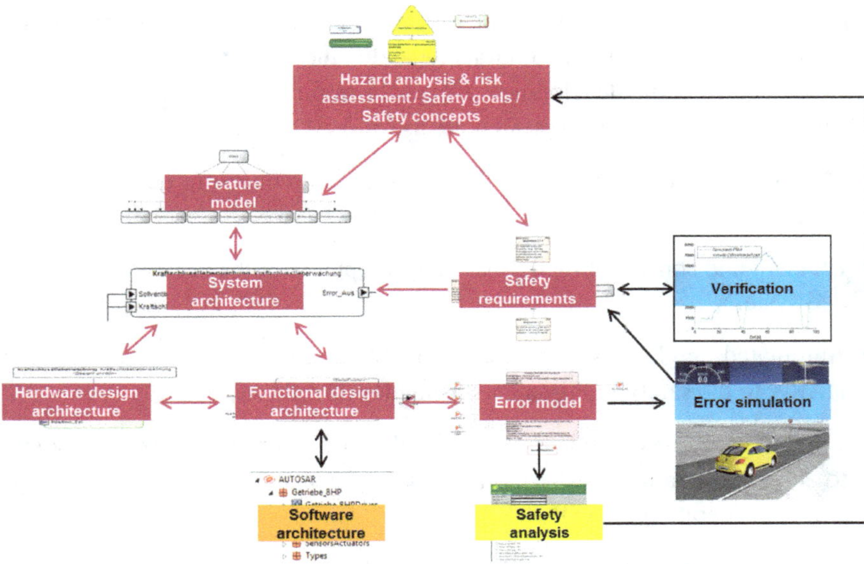

Figure 4.20: Traceability of the development steps

Another major advantage of the developed approach is the mapping of the EAST-ADL levels to the work products and requirements of ISO 26262. Already, in the initial phase, the relationship between EAST-ADL and ISO 26262 is very significant. With the EAST-ADL dependability package, the safety aspects from ISO 26262 can be considered early in the architecture development workflow. Thus, the HARA is performed to determine the safety goals.

Dependability models allow the modeling of the requirements that are necessary to achieve the defined safety goals. Subsequently, the functional and technical safety concept can be created. With these resources, the user can gain an overview of which safety concept is formulated for which safety goal. EAST-ADL offers the possibility to automatically generate the error models from FDA-models. With help of this approach and the support of external analysis tools, the safety analysis can be performed automatically using the error models.

Systematic errors can be detected by the verification of the requirements and error simulation and this method also enables the specification of the safety measures early in the development phase. In summary, the advantages of the resulting method from this research are as follows:

- Modeling of the safety-related functions in an architecture description language.
- Efficient and consistent model-based development of the automotive embedded system.
- Automated, model-based generation of ISO 26262 work products from the HARA to the safety requirements using developed scripts.
- Combination of system and safety development to achieve traceability and to show the relationship between the safety goals and the safety functions with consideration of ASIL.
- Verification and validation of the functional safety concept and the technical safety concept in an earlier project development phase.
- Early detection of systematic errors and the impact of errors on the system.

4.2 A Model-driven Approach for DFA Using Modified EAST-ADL

This subchapter presents an approach for model-based DFA which is required by ISO 26262 for applying decomposition to safety functions. The proposed method involves extending the hardware modeling, function modeling and dependability package of EAST-ADL to permit the modeling of a multicore processor with its hardware elements and the software safety architecture. The modeling of these elements is necessary to prove hardware and software independency. Also presented in this section of the research document are scripts for the automatic analysis of the decomposition paths from the system level to the software and hardware level and for generating the analysis results. The extensions to EAST-ADL and the developed scripts make it possible to gain sufficient transparency and traceability for the safety arguments and to support the entire safety process in a single solution, even during hardware and software development.

4.2.1 Introduction

The ISO 26262 standard permits the application of the decomposition approach to higher safety requirements up to ASIL-D. An appropriate decomposition has the advantage of decreasing the ASIL rating of the top events, but the use of ASIL decomposition also demands the redundancy of the safety requirements and that the redundant requirements be allocated to sufficiently independent architectural elements. For the decomposition approach, ISO 26262 requires proof of DFA to provide evidence of sufficient independence between the decomposed function parts. Currently, it is possible to use commercial external tools to perform only the safety analysis within EAST-ADL. These tools are capable of automatically generating FTA and FMEA from an EAST-ADL error model. But, at this point in time, the dependent failure analysis must be performed manually; there is no tool that supports this analysis. This manual step causes significant additional development effort because the entire path from the system decomposition down to the software and hardware decompositions must be analyzed to ensure that the signals and hardware parts are sufficiently independent. The other disadvantage of the manual analysis is that it is difficult to achieve traceability.

One of the approaches with which the engineers deal with these challenges uses model-driven system, software and safety development. This type of development makes it possible to describe, analyze and verify the system, software and safety architecture with models, in order to detect the design and systematic errors before the implementation or code generation. Due to the high complexity of E/E-Systems in the vehicle, the creation of the system and software architecture is divided into different abstraction levels. This allows the design to be checked on each level, which leads to a very early evaluation of the system architecture based on the functional and non-functional requirements.

4.2.2 DFA According to ISO 26262

4.2.2.1 Approach of System and Safety Modeling

The developed approach described in chapter 4.1 shows how the system, software and safety development are merged, thus decreasing the complexity of the system with a continuous architecture.

This method enables consistency and traceability in the individual steps by permitting the efficient tracing from the software architecture to the feature model and from the safety analysis to the HARA. This new approach for performing a model-based DFA is described in detail in the following text.

4.2.2.2 Requirements for DFA

As described in subsection 2.2.3, the ISO 26262 standard permits the use of the decomposition approach for high level safety requirements. A precise decomposition provides the advantage of decreasing the ASIL rating of the safety goals, but the application of ASIL decomposition requires the redundancy of safety requirements and that these requirements are allocated to sufficiently independent architectural elements.

Further requirements are described in ISO 26262-part9, clause5. As shown in Fig. 4.21, the ASIL of the safety goal is inherited by the corresponding safety requirements and safety functions. Fig. 4.21 also shows that the decomposed functions of sufficiently independent architectural elements inherit the original ASIL information in the brackets.

The most important aspect and requirement of the decomposition approach is to prove the absence of dependent failures. The purpose of the DFA is to find single causes that could prevent a required independence or a freedom from interference requirement between architectural elements from being fulfilled, which constitutes the violation of a safety requirement. As stated in ISO 26262-part9, the independence of architectural elements is threatened by common cause failures and cascading failures, while FFI is only violated by cascading failures as shown in Fig. 4.22.

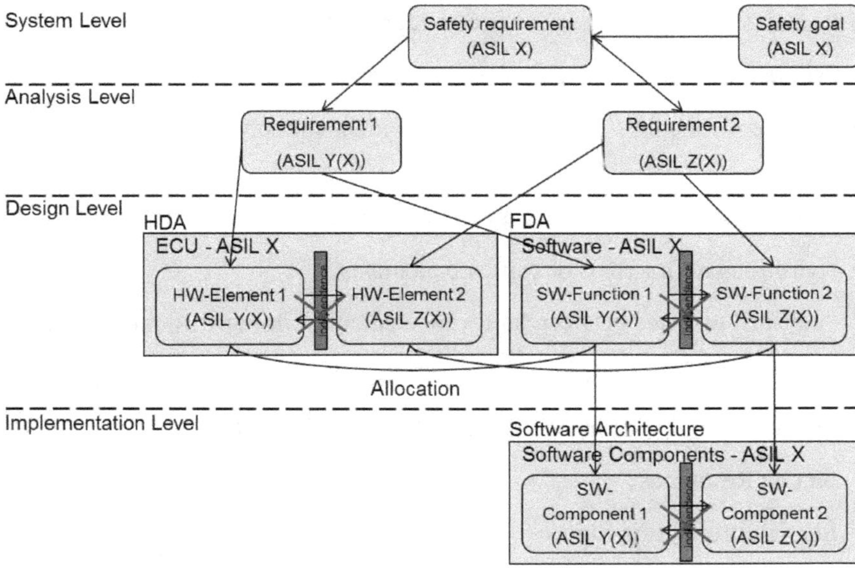

Figure 4.21: ASIL-Decomposition within EAST-ADL

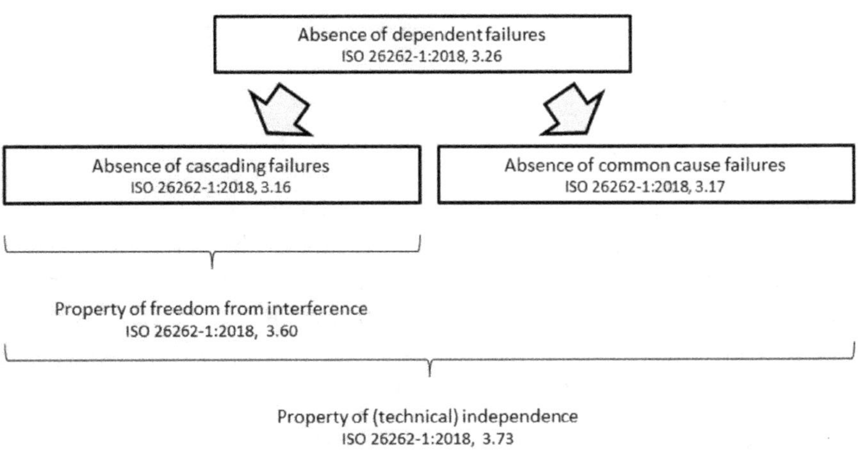

Figure 4.22: Relationship between different classes of dependent failures [14]

The analysis of dependent failures considers architectural features such as [14]:

- "similar and dissimilar redundant elements"
- "different functions implemented with identical software or hardware elements"
- "functions and their respective safety mechanisms"
- "partitions of functions or software elements"
- "physical distance between hardware elements, with or without a barrier"
- "common external resources."

Dependent Failure Initiators according ISO 26262-part9, annexC:

- Shared Resources
- Shared Information Inputs
- Environmental Influences
- Systematic Coupling
- Components of Identical Type
- Communication
- Unintended Impact

As mentioned in subsection 3.4.3, there are additional challenges to consider when performing a DFA for AD systems because of the SOTIF-related performance limitations. The technological shortcomings of the complex sensors and complex algorithms should be considered as additional SOTIF-related dependent failure initiators. Fig. 4.23 illustrates the extended DFA methods with the SOTIF-related technological limitations.

Fig. 4.24 and Fig. 4.25 show how SOTIF triggering events related to the sensors such as weather conditions can influence the decomposed functions and also fail-operational safety design. The independence of architectural elements is threatened by this kind of triggering event which lead to a safety goal violation. These triggering events should be analyzed systematically in order to

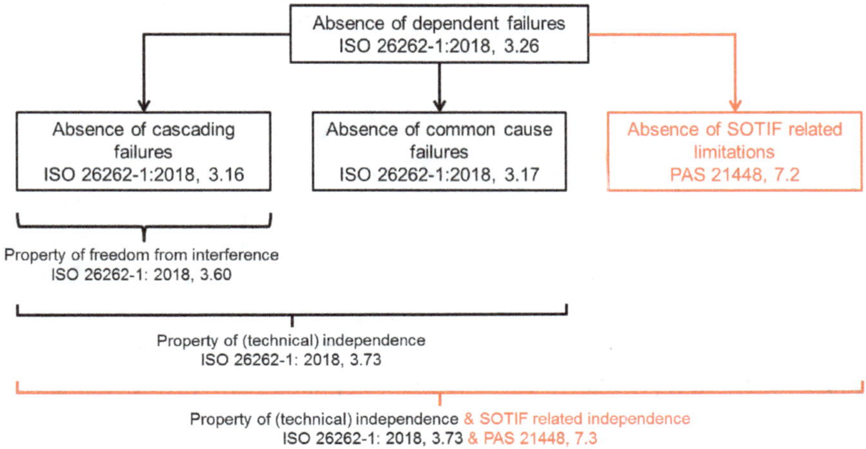

Figure 4.23: Extended DFA

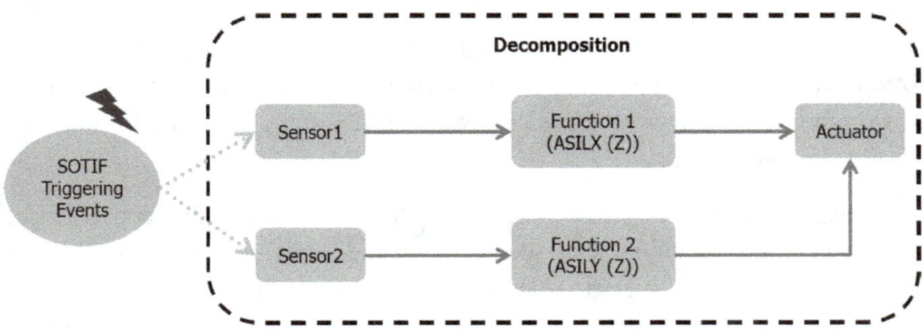

Figure 4.24: SOTIF Dependent Failure Initiators for decomposition

avoid a potentially hazardous system behavior. presents a solution for the automatic analysis of such dependent failure initiators.

This research focuses on the analysis of such architectural features as similar redundant elements, partitions of functions or software elements, shared resources and the used signals. The model-based DFA, which was developed in the context of this research, is able to check whether the evidence of sufficient independence exists between the decomposition paths and signals. The details of this approach are described in the following sections.

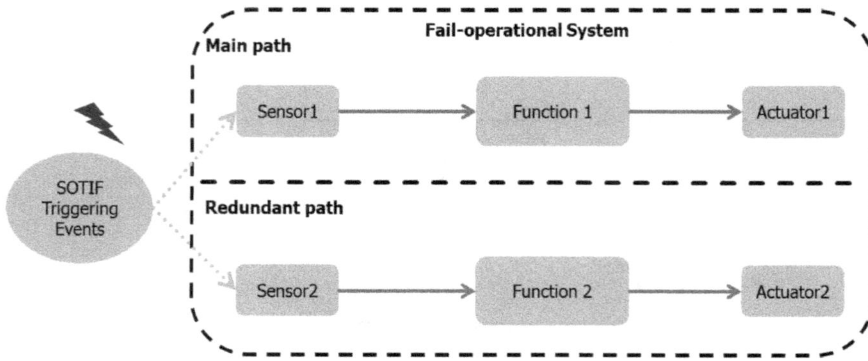

Figure 4.25: SOTIF-related dependent failure initiators for diverse paths

4.2.3 Necessary Developments of EAST-ADL for the DFA

The purpose of EAST-ADL is to provide the engineers with facilitation for the representation and description of electronic systems in vehicles in a standardized form. It can be used for different activities, such as modeling of functional requirements, safety work products from ISO 26262, as well as for analysis and design purposes.

The EAST-ADL metamodel is organized into four different abstraction levels, the "system level, analysis level, design level and implementation level". Each of these levels fulfills specific roles, and each level considers a different phase of the vehicle development and provides a different perspective of the entire system architecture. But currently, the safety aspect is not considered in the modeling of these abstraction levels. The safety modeling is done separately in the dependability package of EAST-ADL, in which the ISO 26262 work products can be modeled. It is necessary to extend the library elements of EAST-ADL with additional safety features which are used for the architecture modeling in the abstraction levels (FAA, FDA, HDA, etc.), so that an automated dependent failure analysis approach can be developed.

Fig. 4.26 shows the extended attributes such as ASIL, type of signals and hardware elements:

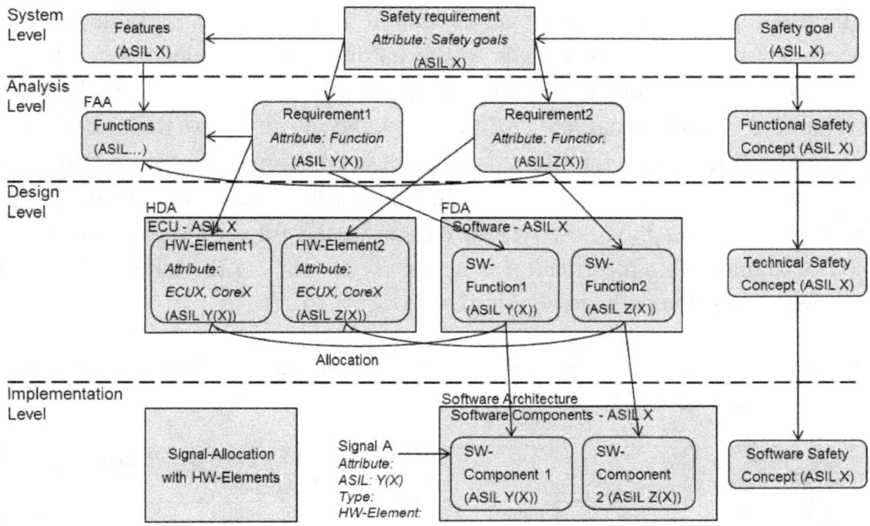

Figure 4.26: Extended attributes for EAST-ADL modeling elements

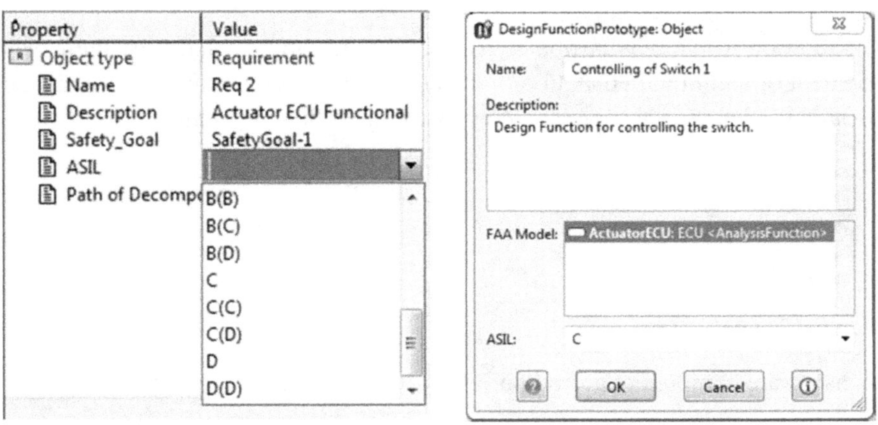

Figure 4.27: Implementation of new features to the EAST-ADL library elements

Fig. 4.27 shows how these new features are implemented for the EAST-ADL library elements.

The development of an additional allocation table as shown in Table 4.1 was necessary to allocate the signals to the hardware library elements. This allocation table contains information about the signals such as the ASIL of the signal, the type of signal, etc. Additionally, the information "Requested from functions" indicates which functions request this signal for the computation and the information "HW-Element" indicates which hardware peripheral elements, e.g., timer modules, ADCs, etc., are used for the preparation of the sensor signal. This information is necessary to analyze the independence of the decomposition paths in a system.

Table 4.1: Allocation table for extended signal features and related hardware elements

	ASIL	Type	HW-Element	Requested from functions	PIN
Sig. A	A/B/C/D	Sensor (Speed)	TIM X, ADC X	Function X	X
Sig. B	A/B/C/D	Sensor (Temperature)	TIM Y, ADC Y	Function Y	Y
Sig. C	A/B/C/D	BUS	TIM Z, ADC Z	Function Z	Z
Sig. X	A/B/C/D	...	...	...	...

In addition to the new attributes for the existing library elements, the development of additional HDA library elements is necessary. Currently, hardware design modeling is abstract and consists of the following elements:

- sensor
- actuator
- node (ECU)
- power supply
- hardware component prototype
- logical bus
- IOHardwarePin
- communication hardware pin
- power hardware pin
- hardware connector

As shown in Fig. 4.28 and 4.29, the HDA library should be developed further with the following elements relevant for multicore processors:

- Analog Digital Converters
- Watchdog
- RAM
- ROM
- GTM unit
- individual cores of manycore processors
- system peripheral bus
- additional global memory elements

Fig. 4.29 shows the library elements developed in this research for HDA modeling. These elements enable the modeling of the hardware architecture of multicore processors and domain ECUs with more processors. It is also possible to model the peripheral elements of the ECU. These models make it possible to analyze the hardware architecture regarding dependent failures and to establish whether the hardware elements in the decomposition paths are sufficiently independent.

Lastly, Table 4.2 shows some triggering events derived from ISO 21448 (SOTIF). These triggering events are allocated to the extended sensors as shown in the table and then each sensor is evaluated for technological insufficiencies relevant to these triggering events. The letter Y is for Yes and N for No.

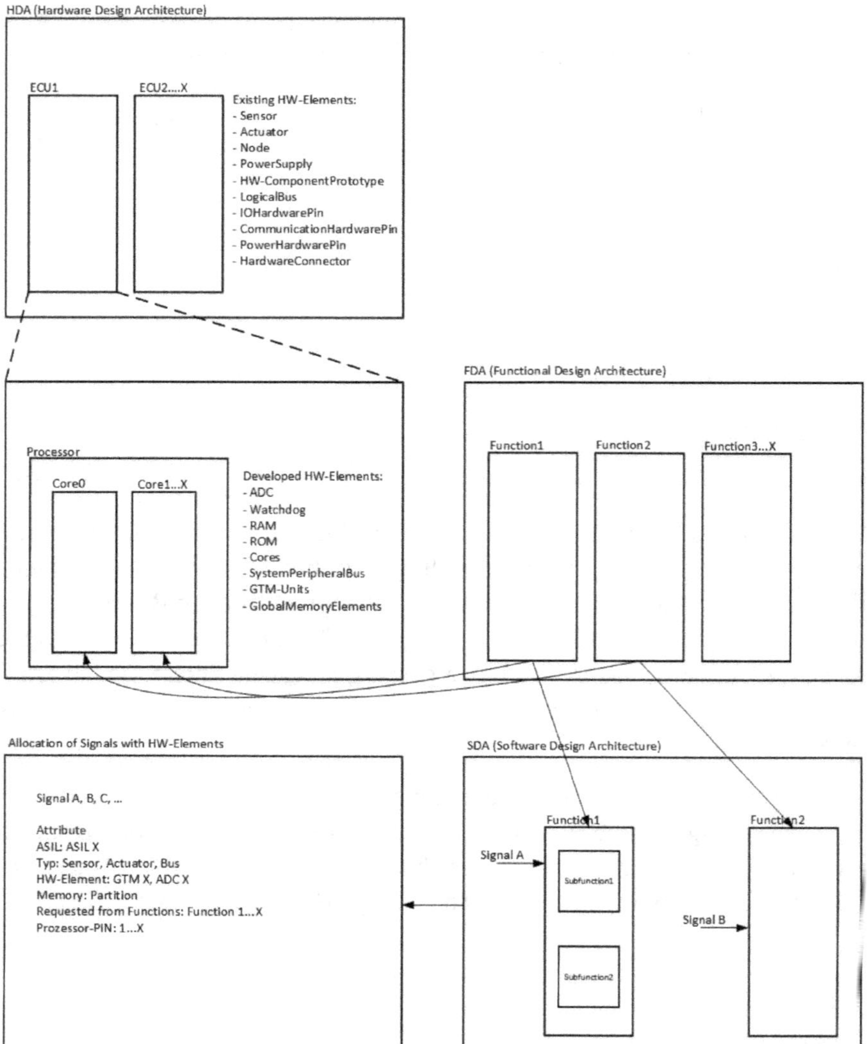

Figure 4.28: Developed hardware library elements considering multicore processors

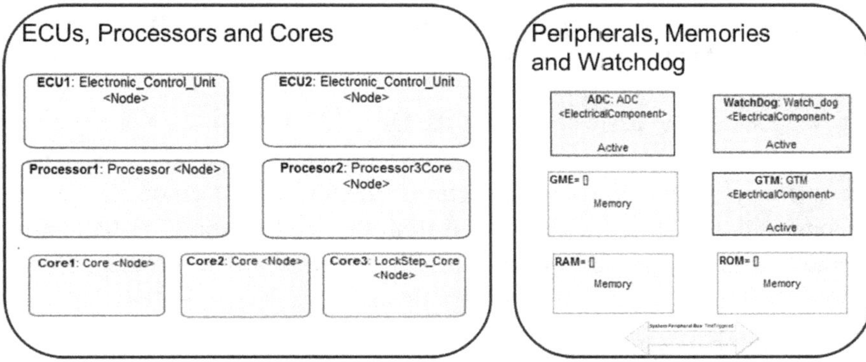

Figure 4.29: Developed hardware library elements

Table 4.2: SOTIF Triggering Events

SOTIF Safety Analysis Method						
Types	**Factors**	**Relevant for UseCases**	**Sensor1**	**Sensor2**	**Sensor3**	**Sensor X**
climate	fine					
	cloudy					
	rainy	Y	N	Y	N	N
	sleet					
	snow (accumulation of snow)					
	hail					
	fog					
time of day	early morning					
	daytime					
	evening					
	night time					
shape of road/lane	straight					
	curve					
	downhill					
	uphill					
	banked road					
	step difference					
	uneven spot (uneven road)					
	Belgian brick road					
	narrow road					
	wide road					
XYZ	...					
	...					

4.2.4 Description of Developed Model-based Approach for DFA and Safety Analysis

Fig. 4.30 shows the basic concept of the DFA. The relationships of the development steps and decomposition paths are modeled within EAST-ADL. There are two ways to analyze the architecture. The top-down method enables the decomposition path to be determined from the model data. The bottom-up method is used to provide the evidence that there is sufficient independence between the architectural elements and signals.

As shown in Fig. 4.30, the system modeling and safety modeling are merged together. It is necessary to extend the dependability modeling in several ways. Additional attributes are necessary for the safety goal definition, the HDA requires enhancement regarding multicore processor elements and the SDA is extended with a HW element allocation table for the signals. This allocation table contains the signals, which are used for the safety-related functions, and the signal attributes such as ASIL classification, type etc. There is also additional information about which processor pin belongs to which signal, and about the signal processing, i.e. which analog digital converter (ADC) and timer modules (GTM, CCU6, GPT12, etc.) are used to create the signal. These extensions and information are necessary in order to perform the DFA automatically and in order to provide the evidence of sufficient independency between decomposed function parts.

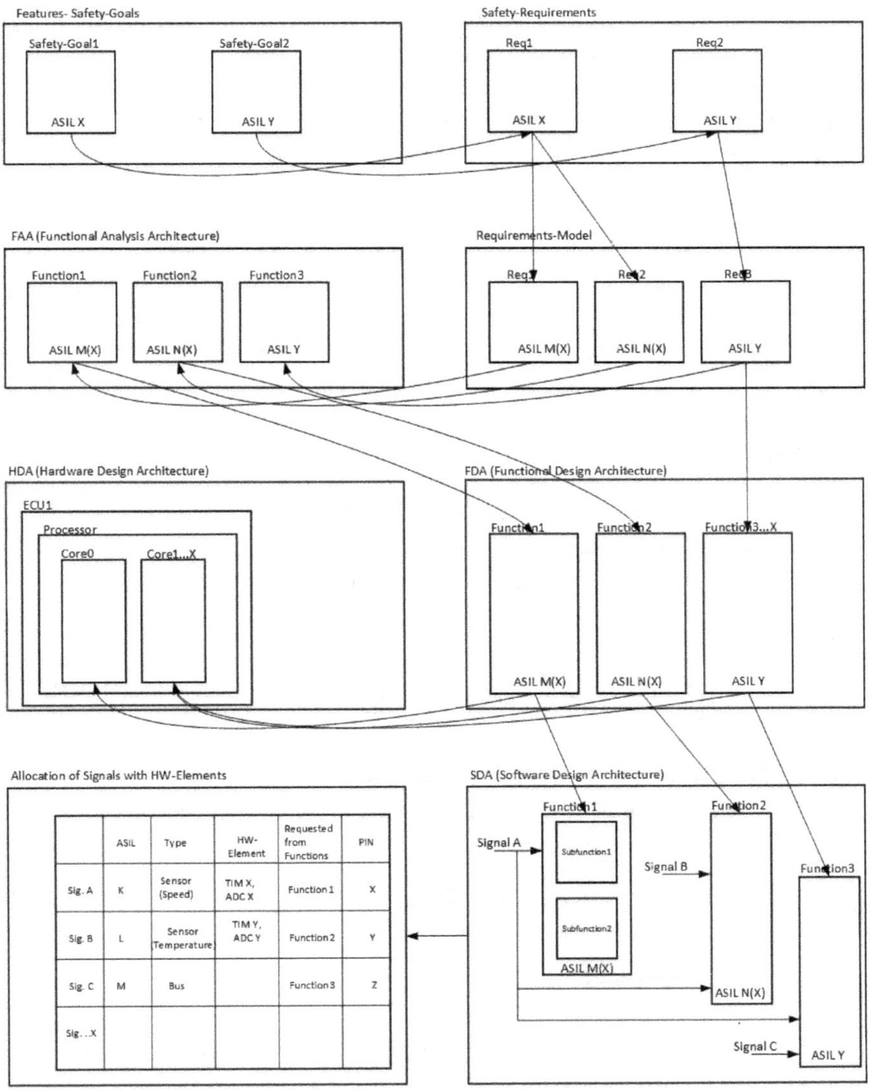

Figure 4.30: Model based DFA

This method enables the traceability over the general system model and shows the relationship between the system and the safety architecture. Hence, it is easier to determine which safety goal and safety requirements are implemented in which software safety function, and which software safety function runs on which ECU and core. The signals are also traceable to the requirements, functions and HW-Elements.

The following DFA rules are developed for the model check to prove the independence of the decomposition paths and to analyze the safety concept compliancy:

1. **Relation Check:**
   a Safety-Goal1 – > Req1 (Safety-Requirements) Safety-Goal2 – > Req2 (Safety-Requirements)

   b Req1 (Safety-Requirements) – > Req1 (Requirements-Model) – > Req2 (Requirements Model)
   Decomposition (one-to-many relationship)

2. **ASIL Check:**
   a ASIL of Safety-Goal1 = ASIL of Req1 (Safety-Requirements) ASIL of Safety-Goal2 = ASIL of Req2 (Safety-Requirements)

   b ASIL of Req1 (Safety-Requirements) = ASIL (in brackets) of Req1 and Req2 (Requirements-Model) ASIL of Req2 (Safety-Requirements) = ASIL of Req3 (Requirements-Model)

   c ASIL of Function1 (FAA) = ASIL of Req1 (Requirements-Model) ASIL of Function2 (FAA) = ASIL of Req2 (Requirements-Model) ASIL of Function3 (FAA) = ASIL of Req3 (Requirements-Model)

   d ASIL of Function1 (FAA) = ASIL of Function1 (FDA) ASIL of Function2 (FAA) = ASIL of Function2 (FDA) ASIL of Function3 (FAA) = ASIL of Function3 (FDA)

   e ASIL of Function1 (FDA) = ASIL of Function1 (SDA) ASIL of Function2 (FDA) = ASIL of Function2 (SDA) ASIL of Function3 (FDA) = ASIL of Function3 (SDA)

3. **Independency Check:**

   a Core (HDA) of Function1 (FDA) != Core (HDA) of Function 2 (FDA)

   b Core (HDA) of Function3 (FDA) can be the same as the core of other functions, because the Function3 is not part of the decomposition.

4. **Signal Check:**

   a ASIL of Signal A >= ASIL of Function1 (SDA) ASIL of Signal B >= ASIL of Function2 (SDA) ASIL of Signal C >= ASIL of Function3 (SDA) ASIL of Signal A >= ASIL of Function3 (SDA)

   b TIM and ADC of Signal A != TIM and ADC of Signal B

   c Input signals of Function1 != Input signals of Function 2

   d Output signals of Function 1 != Input signals of Function 2

5. **SOTIF Safety Analysis Check**

   a If (Relevant use cases **contain** Triggering Events Factor(i)) &
   (Triggering Events Factor(i) **affects** (Sensor1 & Sensor2))
   Then
   Decomposition violation

   NOTE: i = Number of triggering events (weather conditions, road conditions, etc.)
   NOTE: Sensor1 and Sensor2 are in decomposition relation or main/ redundant path relation.

In order to facilitate the understanding of the rules, the effects of the rules can be visualized in the following architecture model:

1. **Relation Check:**

   The rules of the relation check verify whether the safety goals are allocated to the correct functional and technical safety requirements, as shown in Fig. 4.31. In the case of decomposition, the safety goal is broken down to the independent safety requirements. These safety requirements must be able to achieve the safety goal independently. For example, if one diagnostic function of a decomposition path is not able to achieve the safety goal because of a failure, the second independent diagnostic function must be able to prevent the safety goal violation.

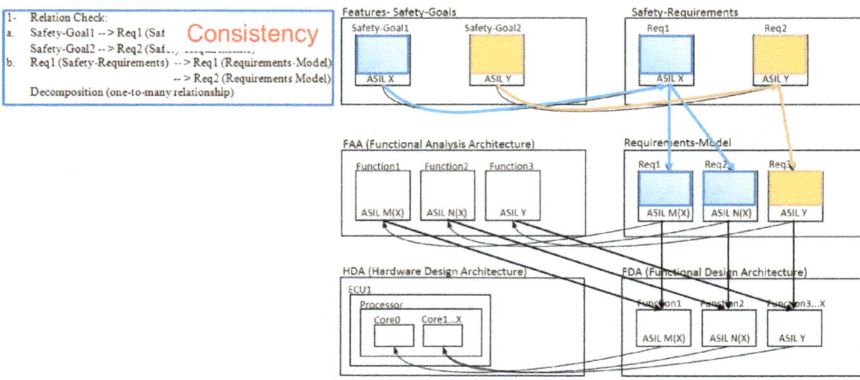

Figure 4.31: Relation Check: Verification of the relationship between safety goals and safety requirements

2. **ASIL Check:**

   The relation check rules confirm whether the ASIL of the safety goal is inherited correctly by the corresponding functional safety requirements, technical safety requirements and functions, as shown in Fig. 4.32. The functions must be implemented according to the ASIL of the safety goal. In the case of decomposition, the independent paths receive the suitable ASIL according to the ISO 26262-part9, clause5.4.9. In such a case, it is necessary to indicate the original ASIL of the safety goal in brackets. In the check of the ASIL, this property is also considered to have the correct decomposition ASIL.

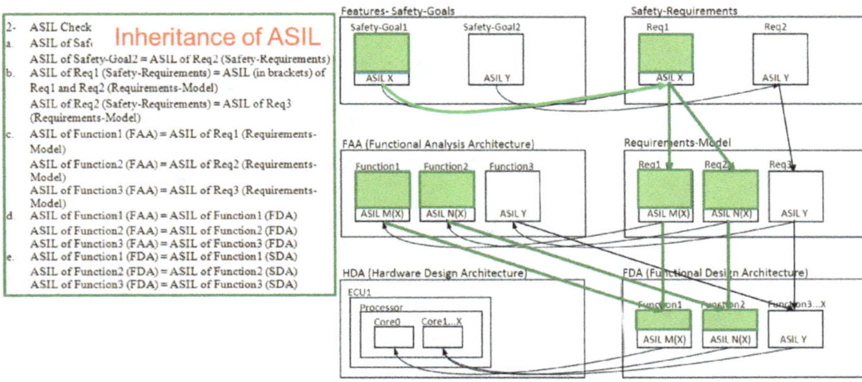

Figure 4.32: ASIL Check: Verification of the ASIL inheritance from safety goals to the safety requirements and safety functions

3. **Independency Check:**

   The rules also confirm whether the decomposed functions are allocated to sufficiently independent hardware elements, as shown in Fig. 4.33. It is necessary for differing processors or cores to perform these functions. The other functions, which are not a part of the decomposition, can be allocated to the same cores, but with consideration of the freedom from interference requirements according to ISO 26262-part6, annexD.

4. **Signal Check:**

   The relation check rules verify, on the one hand, whether the signals of the decomposed functions are independent of each other, and on the other hand, whether the ASIL level classification of the signal is appropriate for the functions, as shown in Fig. 4.34. The signals should have at least the same ASIL level as the ASIL level of the function. The differing inputs for the decomposed functions are required to prevent cascading failures. If the same signal is used by both decomposition functions, this can lead to a safety goal violation. This is also applicable to the relationship of the functions to each other. For example, the output signal of a decomposed function is not permitted to be used for the other decomposition path.

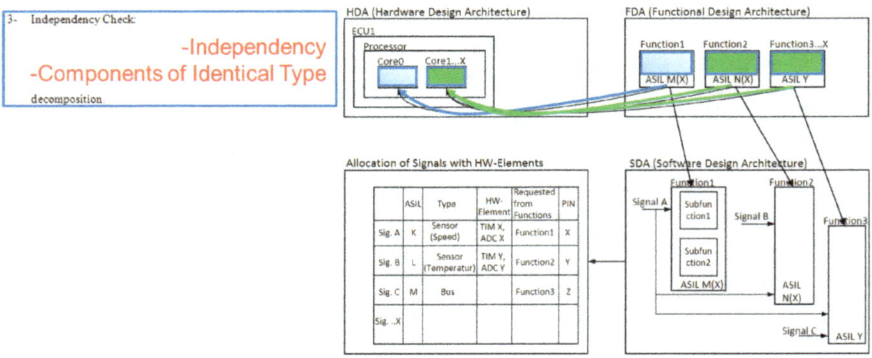

Figure 4.33: Independency Check: Verification of the independence of the decomposed functions

Additionally, the sensor signal conditioning for signals which are used for the decomposed functions should also be done independently. Using the same hardware peripheral elements for the conditioning of the signals, e.g., the ADC, the timer modules, etc., is also not permitted.

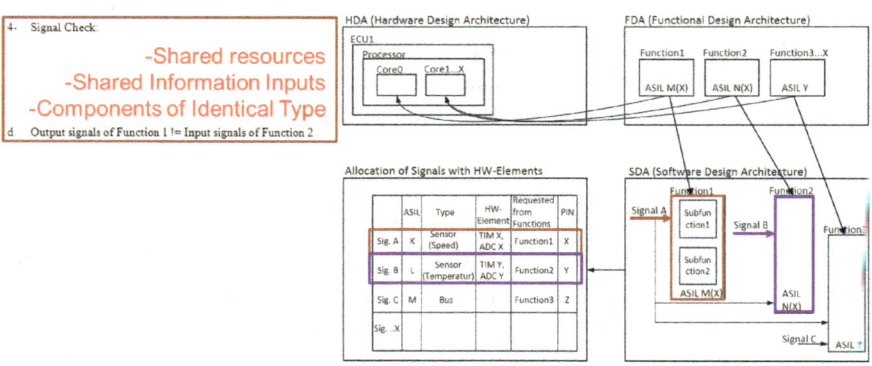

Figure 4.34: Signal Check: Verification of signal independence and quality

5. **SOTIF Safety Analysis Check:**

   The rules verify whether a triggering event affects the sensor signals and related perception algorithms which are applied for decomposed functions or different paths of diverse design. The decomposed functions should not be simultaneously affected by the triggering events. Otherwise, one or more triggering events can violate these independent functions, which can lead to a safety goal violation. The Fig. 4.35 illustrates the relationship between the use cases and triggering events.

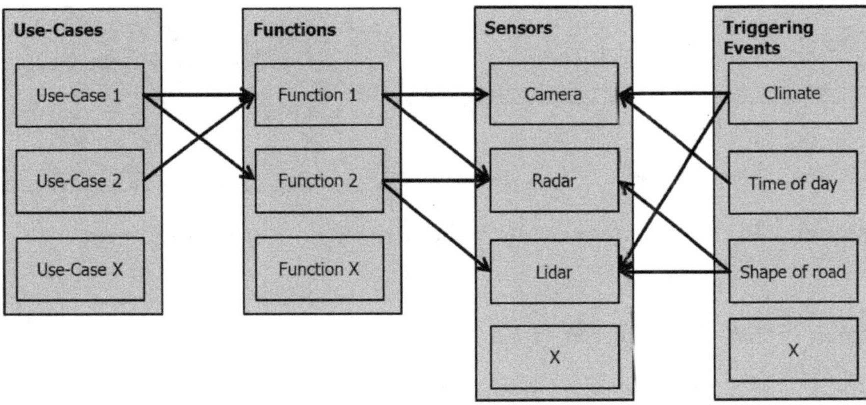

Figure 4.35: Relationship between use cases and SOTIF triggering events

The Fig. 4.36 shows how a triggering event such as heavy rain or snow can affect the different sensor signals and also functions.

Thus, these rules are able to check the following dependent failure initiators that are listed in subsection 4.2.2.2:

- Shared Resources
- Shared Information Inputs
- Components of Identical Type
- Communication
- SOTIF triggering events

The next chapter shows the tool developed for the DFA which contains the rules to check the independency. The result of the analysis is also documented in the same tool.

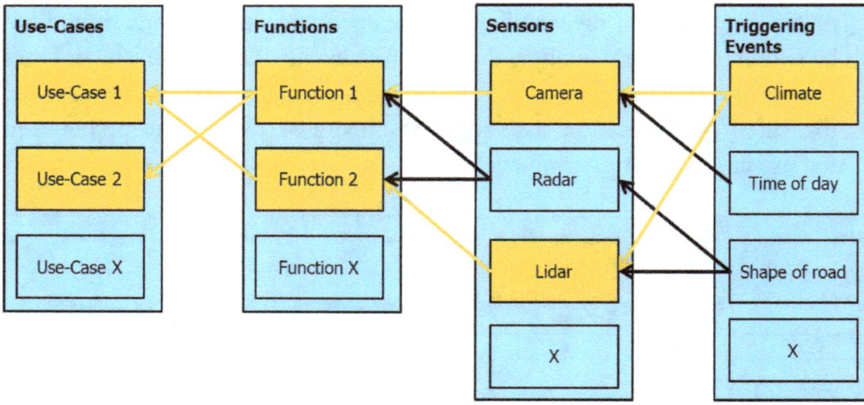

Figure 4.36: SOTIF safety analysis check considering triggering events

4.2.5 Scripts and Reports

Fig. 4.37 shows the graphical user interface (GUI) for the developed DFA. This GUI contains the DFA checks and allows the scripts to be performed automatically. The DFA rules are implemented as shown below in four different scripts, namely, the relation check, the ASIL check, the independency check and the signal check. The scripts can be performed independently and also for all of the EAST-ADL abstraction layers, for example, FAA model, FDA model, dependability model, requirements model, etc.

It is possible to obtain the results for each execution of the rules, so it is easier to provide the DFA status. If the status is marked red, the user can investigate the violation directly in the output files. For each rule, the status can be found directly at the bottom of the rule as “Check is Ok” or “Check is not Ok”. This is shown in the following Fig. 4.38. The status display helps to discover the violations quickly and to correct them.

The following example in Fig. 4.39 shows the relationship between safety goals, safety requirements and safety functions. After the analysis, the decomposition path and the additional corresponding elements are marked with a color to make the identification of the independence path easier. Another advantage of this representation is that it assists in the recognition of the decomposition paths. For example, the provision of a new signal protects against a violation of the independency rules. The signals are allocated to hardware elements such as TIMs and ADCs in order to analyze the hardware independence.

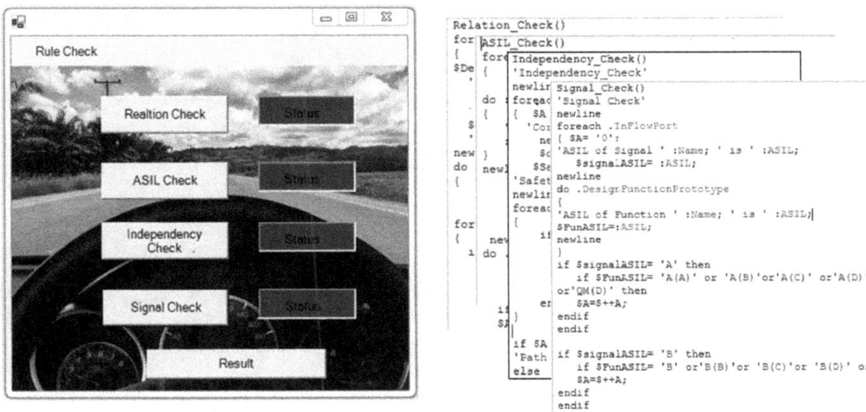

Figure 4.37: GUI and Scripts for DFA check

In this case, the signal independence is violated by the decomposition path.

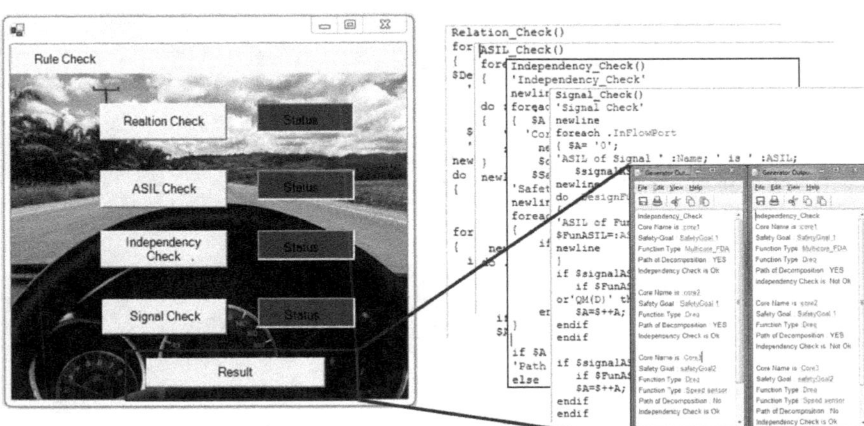

Figure 4.38: DFA check and results

Signal A in Fig. 4.39 is requested by two different functions which are part of the decomposition and therefore should be independent of each other.

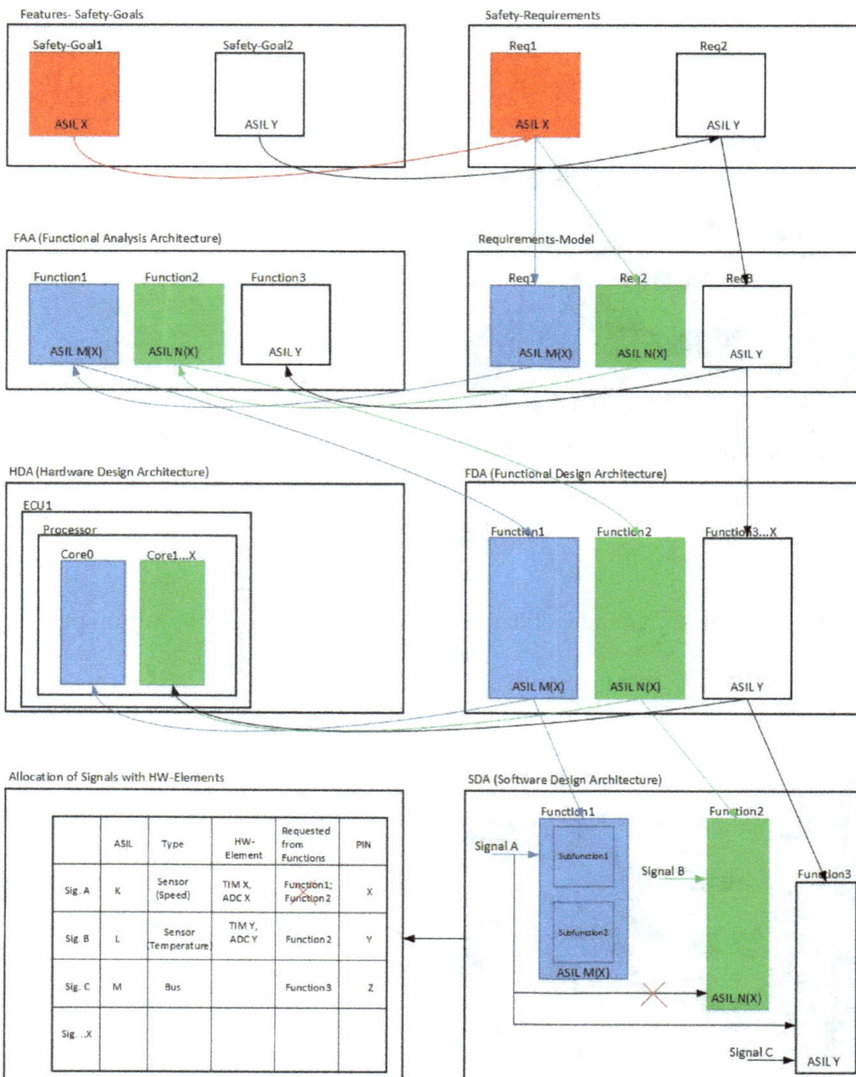

Figure 4.39: Use case and reports

4.2.6 Use case

The developed model-based and automatized DFA is evaluated in the following example shown in Fig. 4.40. This example illustrates how the safety goal can be achieved with independently decomposed safety requirements. This figure is extracted from the ISO 26262–part10, clause11.3 and extended with an additional external sensor. In this example, the actuator is controlled by an actuator control ECU. This ECU receives the information of a driver request to open the door automatically. This driver request is diagnosed in the actuator control ECU with regard to the vehicle speed signals transmitted over the BUS from an external vehicle ECU "VS ECU" and a speed signal from the sensor. If the vehicle speed signal value from the VS ECU is greater than a defined threshold, then the switch remains in the open position and the current is not forwarded to the actuator, so the door is not opened. Independently, the redundant switch control is based on the redundant speed sensor signal. If the speed sensor signal value is less than a defined threshold, then the redundant switch is in the "closed" position, otherwise, the switch remains open.

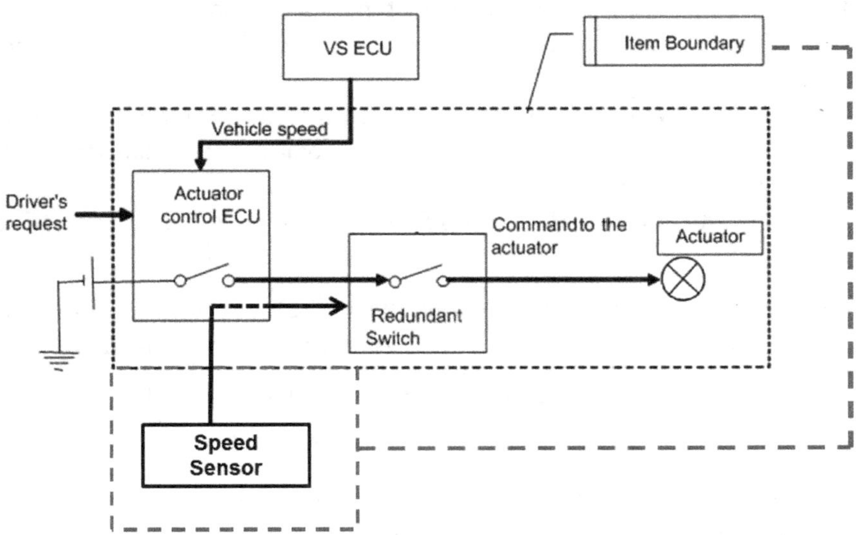

Figure 4.40: Use case: Extended item boundary

Summarized, the vehicle door is not permitted to open as a result of a driver request, if the vehicle speed is greater than a defined threshold. To prevent the faulty control of the door (unintended opening of the door) based on only one source of vehicle speed information, a redundant path is used for the door opening functionality.

The automatic generation of the DFA is based on the EAST-ADL architecture models. In order to evaluate this approach, this example is modeled below:

4.2.6.1 Dependability Model / HARA:

As described before, the EAST-ADL dependability model is used to create the ISO 26262 safety work products. Fig. 4.41 shows the dependability model for this example. This model contains the safety work products, item definition, HARA, safety goal and functional and technical safety concept. The individual work steps for this example are described below:

Item: Door controlling using door control unit
Hazard: Unintended opening / release of the vehicle doors
Hazardous Event: Unintended opening of the vehicle doors at high vehicle speed.
Safety Goal: Avoid opening the door by preventing the powering of the actuator while vehicle speed is greater than a calibratable speed threshold. This safety goal is classified with ASIL C as shown below.
ASIL-Classification for Hazardous Event: The hazardous event is classified as follows:

- Controllability (C2): Normally controllable (Between 90% and 99% of all drivers or other traffic participants are usually able to avoid harm)
- Severity (S3): Life-threatening injuries (survival uncertain), fatal injuries
- Exposure (E4): High probability

These values result in ASIL C rating. The classification for this example does not reflect the real evaluation of the ASIL and is used only for the illustration of ASIL decomposition.

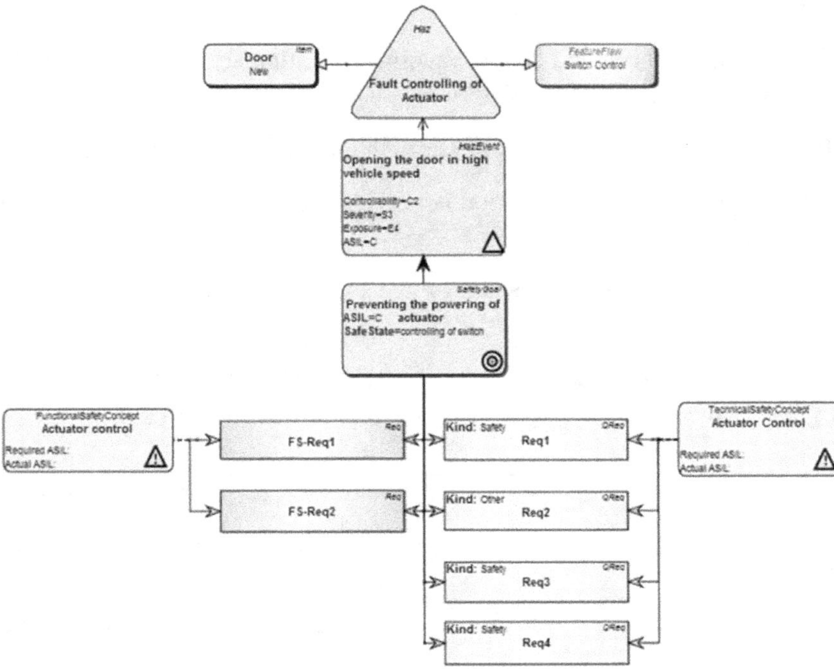

Figure 4.41: Use case: HARA

4.2.6.2 Requirements Model:

Requirements model contains functional safety requirements as well as technical safety requirements. The functional safety requirements are listed below with their ASIL classification in brackets:

- FS-Req1: AC ECU does not power the switch if the vehicle speed via BUS is greater than 15 km/h (ASIL A)
- FS-Req2: Switch is in the open state, if the sensor speed is greater than 15 km/h (ASIL B)

The technical safety requirements are listed below:

- Req1: Vehicle speed and driver request information are provided to the AC ECU from the BUS (ASIL A)
- Req2: AC ECU only provides power with the driver request (ASIL A)
- Req3: Vehicle speed information is provided to the actuator ECU from the speed sensor (ASIL B)
- Req4: Switch is in the open state, if the speed information is not available (ASIL B)

As shown in Fig. 4.42, the safety goal is achieved through two independent functional safety requirements.

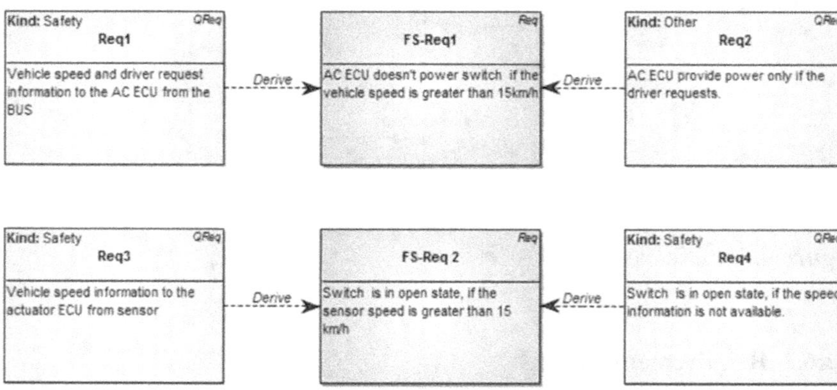

Figure 4.42: Use case: Requirements model

4.2.6.3 Functional Analysis Architecture (FAA):

The FAA model describes the functional analysis architecture at a high level of abstraction. The vehicle speed sensor signal and the actuator are realized with the library element "FunctionalDevice". There are two functions to control the switch. The energizing of the switch (Powering of switch) is performed based on the vehicle speed information from the BUS and the controlling of the switch (Controlling of Switch) is based on the speed sensor signal information, as shown in Fig. 4.43.

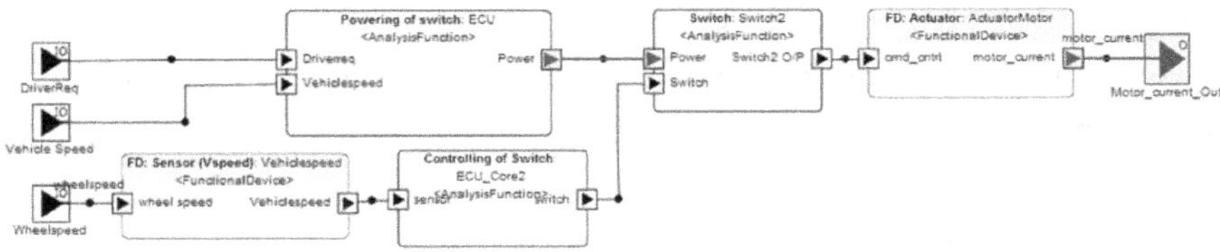

Figure 4.43: Use case: Functional analysis architecture (FAA)

4.2.6.4 Functional Design Architecture (FDA):

The abstract functions of the FAA model are detailed in the FDA model, as shown in Fig. 4.44. In particular, the functional devices of the sensor and the actuator are distributed into three function categories, namely, the hardware function (HardwareFunction), the basic software function (BasicSoftwareFunction) and the local device manager (LocalDeviceManager). The functional safety requirements are implemented using the independent functions which permit the safety goal to be fulfilled.

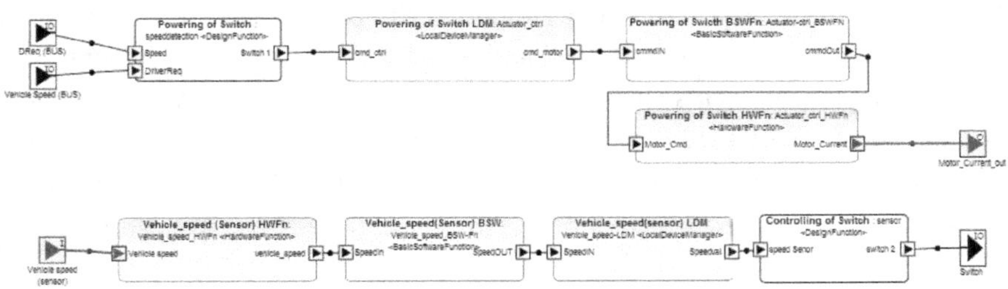

Figure 4.44: Use case: Functional design architecture (FDA)

4.2.6.5 Hardware Design Architecture (HDA):

Fig. 4.45 shows the Hardware Design Architecture. In this example, there is one speed sensor signal designed with the hardware library element "Sensor". Two signals "vehicle speed" and "driver request" are provided via BUS from the different control units.

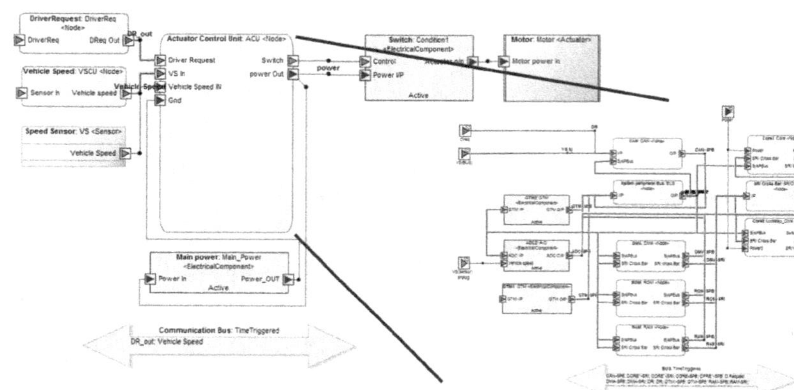

Figure 4.45: Use case: Hardware design architecture (HDA)

The content of the ECU "Actuator Control Unit" is illustrated in Fig. 4.46. In this example, the ECU contains a dual core processor, memory elements and other peripheral elements such as ADCs, timer modules, DMA, communication paths, etc. This hardware design architecture is used to provide evidence of whether the decomposed functions are allocated to sufficiently independent hardware elements.

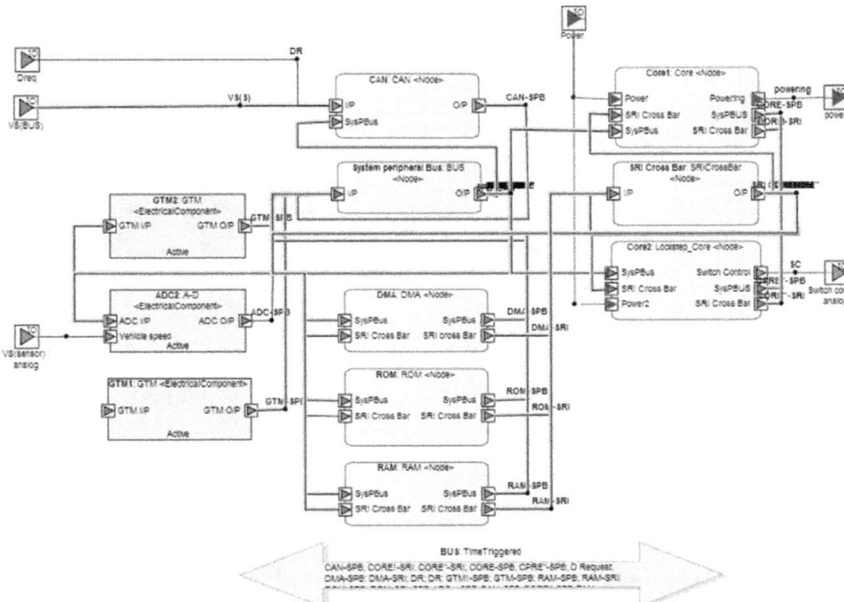

Figure 4.46: Use case: HDA for actuator control unit

4.2.6.6 FDA/HDA Allocation:

The following figures show the allocation of functions from the FDA model to the hardware design elements from the HDA model. The allocation can either be done graphically as in Fig. 4.47 or by using a table as in Fig. 4.48. The functions of decomposition path in ASIL A regarding energizing the actuator via switch are allocated to the core 1 of the dual core processor. The functions of decomposition path in ASIL B regarding the controlling of the switch are allocated to the other core of the dual core processor.

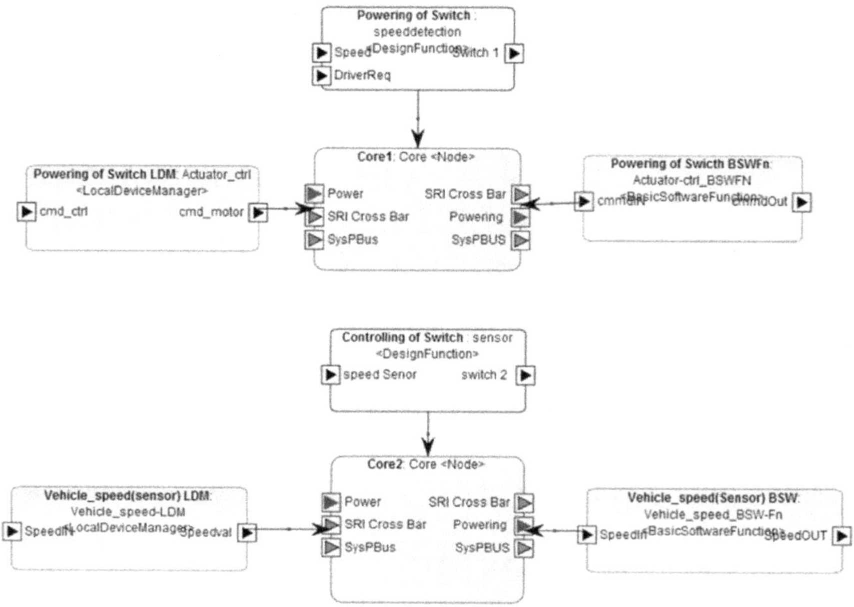

Figure 4.47: Use case: FDA-HDA Allocation in graphic form

	Core1	Core2
Controlling of Switch : sensor <l		FunctionA
Powering of Swicth BSWFn: Ac	FunctionA	
Powering of Switch : speeddete	FunctionA	
Powering of Switch LDM: Actua	FunctionA	
Vehicle_speed(Sensor) BSW: V		FunctionA
Vehicle_speed(sensor) LDM: Ve		FunctionA

Figure 4.48: Use case: FDA-HDA Allocation in tabular form

4.2.6.7 DFA Checks: Relation Check – Rule Violations:

As described before, the relation check verifies whether the safety goals are allocated to the correct functional and technical safety requirements. The script automatically checks the relationship between the functional and technical safety requirements. As shown in Fig. 4.49, the functional safety requirements belong to the same safety goal. This means that both FS-Req1 and FS-Req2 are a part of the decomposition and each can achieve the safety goal independently. On the one hand, FS-Req1 prevents the energizing of the actuator based on the vehicle speed information via BUS. On the other hand, FS-Req2 prevents the powering of the actuator on the basis of the sensor speed signal information. In this case, the result of the relation check, as shown with the generator output in Fig. 4.49, confirms the decomposition because the requirements of the decomposition paths are sufficiently independent.

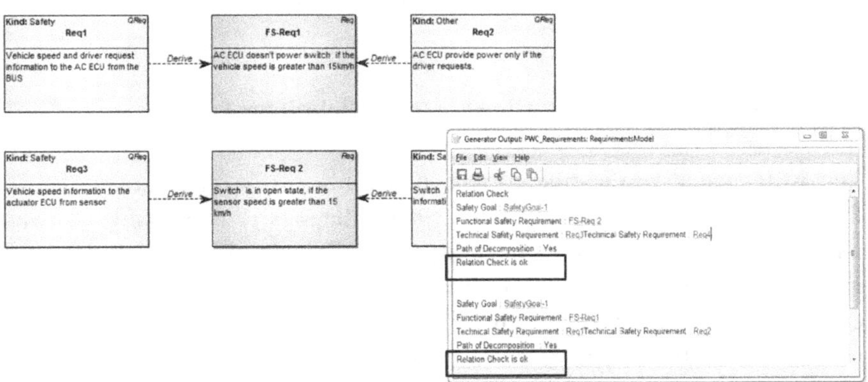

Figure 4.49: Use case: Relation check-Ok/Safety requirements are independent

In contrast, the relation check in the following example does not verify the decomposition, because the technical safety requirement Req1 is used for both functional safety requirements FS-Req1 and FS-Req2. This is not permitted because the same technical safety requirement can affect both functional safety requirements, which leads to a decomposition violation. The same technical safety requirement is only permitted to be used if the ASIL level of the requirement is same as the original ASIL of the safety goal. The scripts are available for the detection of such violations as shown in Fig 4.50.

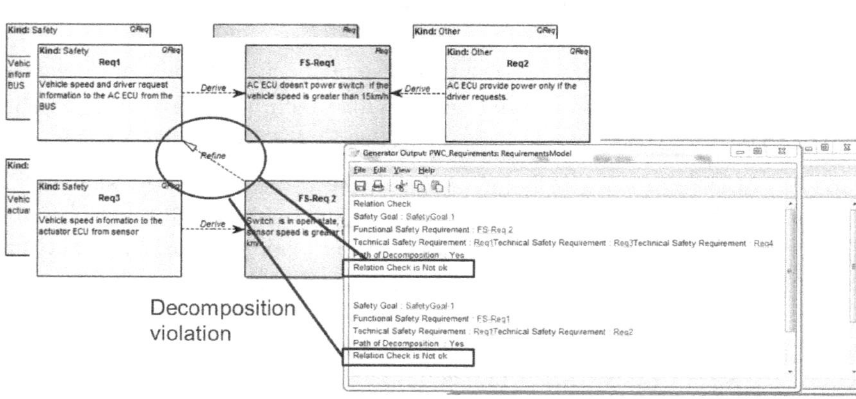

Figure 4.50: Use case: Relation check-Decomposition violation

4.2.6.8 DFA Checks: ASIL Check – Rule Violations:

The ASIL check verifies the correct inheritance of the safety goal ASIL to the corresponding safety requirements and functions. Subsequently, the various architecture models “Dependability model, Requirements model, FAA and FDA” are automatically checked, using the ASIL scripts, to ensure the ASIL consistency for decomposed paths. In this use case, the safety goal in ASIL C is met with the two decomposed safety requirements in ASIL B(C) and A(C). The safety requirements are implemented with the corresponding decomposition functions “Powering of switch” in ASIL A(C) and “Controlling of switch” in ASIL B(C). Fig. 4.51 shows how the first part of the decomposition requirement in ASIL A(C) is correctly inherited from the dependability model to the requirements model, FAA and FDA.

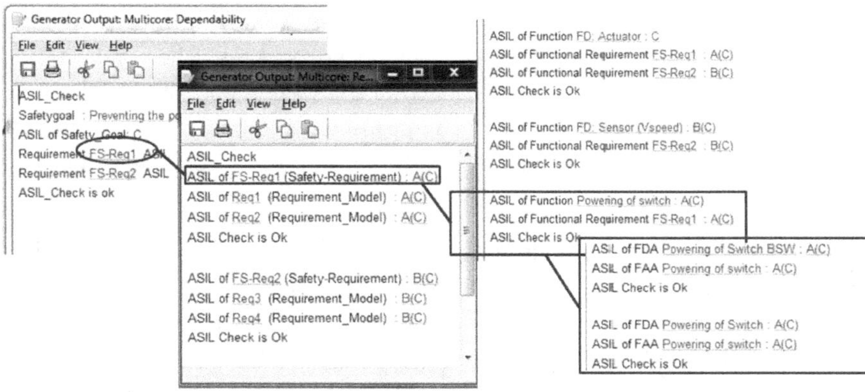

Figure 4.51: Use case: ASIL check for ASIL inheritance

4.2.6.9 DFA Checks: Independency Check – Rule Violations:

This script for the independency check confirms whether the decomposed functions are allocated to sufficiently independent hardware elements.

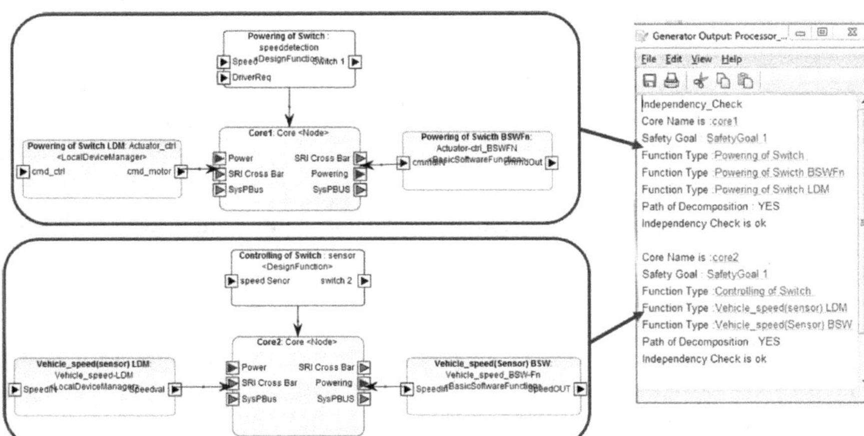

Figure 4.52: Use case: Independency check-Ok/Allocation to the independent hardware elements

As shown in Fig. 4.52, the decomposed functions “Powering of Switch” and “Controlling of Switch” are allocated to independent cores. The cores and the functions allocated to each core are listed in the DFA report, which also contains the analysis results regarding decomposition violations. In this case, there is no violation, because the functions are allocated to independent processor cores.

For purposes of illustration and in order to check the independency violation, the decomposed function “Powering of Switch” is additionally allocated to core 2. In this case, Fig. 4.53 shows that there is a decomposition violation with regard to the independency check. Core 2 contains both decomposition functions, and both functions belong to the same safety goal. This is not permitted and leads to a violation.

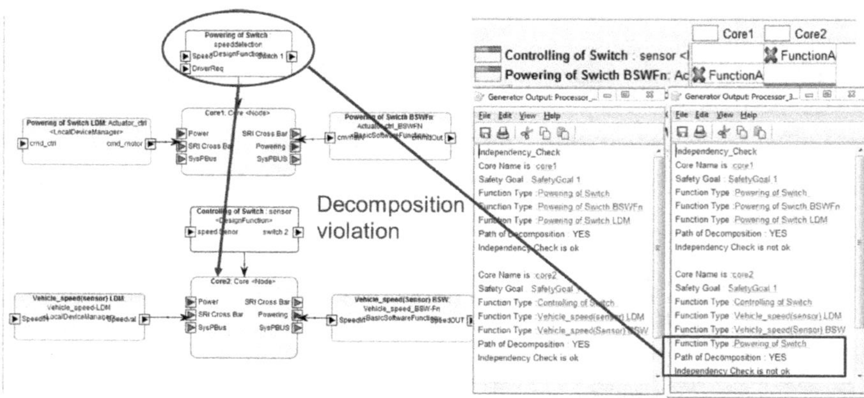

Figure 4.53: Use case: Independency check-nOk/Allocation of decomposed functions is not independent

Fig. 4.54 shows that there is no independency violation, although the same function is allocated to the core 2. This is correct, because the functions do not have a decomposition relationship. In this case, allocation of the signal to the same cores is permitted under consideration of the freedom from interference requirements in ISO 26262-part6, annex C.

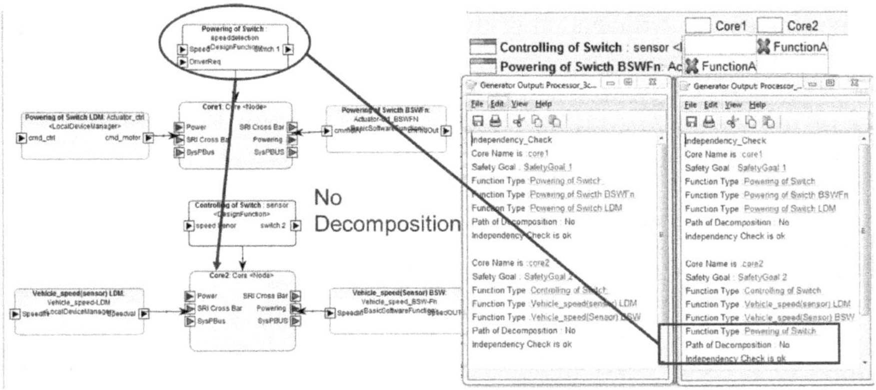

Figure 4.54: Use case: Independency check-Ok/Dependence is permitted because there is no decomposition relationship

4.2.6.10 DFA Checks: Signal Check – Rule Violations:

The signal check script verifies whether

- the signals of the decomposed functions are independent of each other.
- the ASIL level of the signals is plausible to the ASIL level of the function. The signals should have at least the same ASIL level as the function ASIL level.

As shown in Fig. 4.55, the generated report lists the input signals of the functions with the ASIL information. So, there is an automatic check of whether the ASIL of the signals matches the ASIL of the function. On the left-hand side, the ASILs are identical and therefore the result is Ok. But on the right-hand side, the vehicle speed signal via BUS is requested with QM for a function with ASIL A. This violation is determined and documented in the generated report.

The Fig. 4.56 shows that there is no DFA violation if the vehicle speed signal via BUS is used for both functions, because the functions are not in a decomposition relationship. In this case, the same signal can be used for different functions under consideration of the rules for freedom from interference according to ISO 26262.

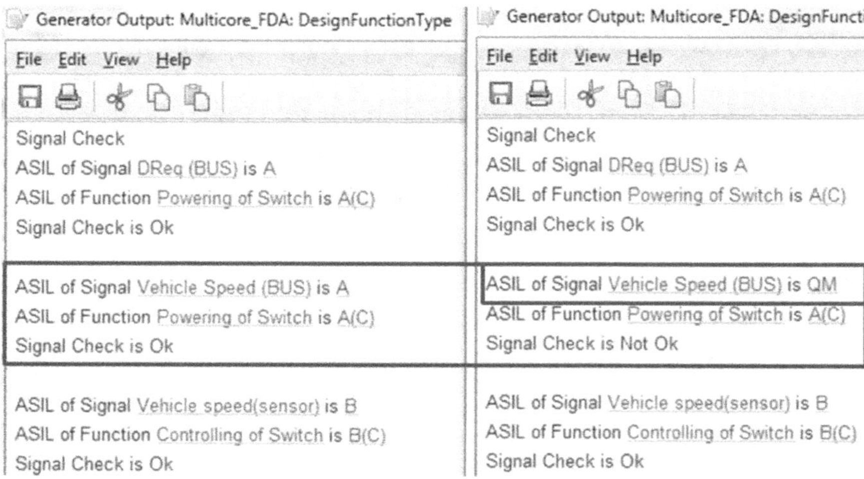

Figure 4.55: Use case: Signal check/ASIL check between signals and related functions

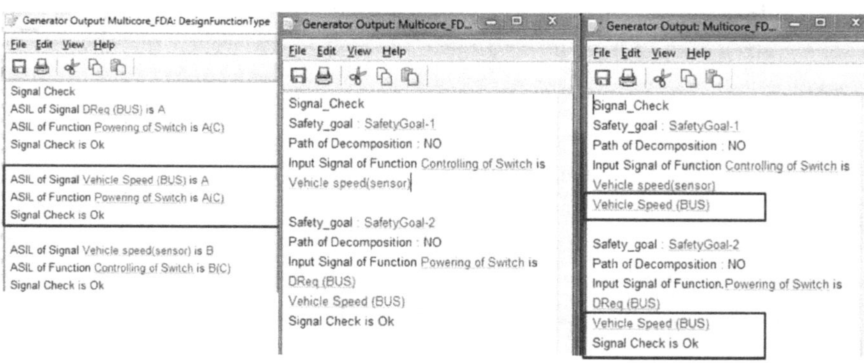

Figure 4.56: Use case: Signal check-Ok/Signal dependence is permitted because there is no decomposition relationship

The following example illustrates a DFA violation, as shown in Fig. 4.57. As shown on the right side of the figure, the vehicle speed via BUS is used for both decomposition functions, which lead to a DFA violation. This signal can affect both functions at the same time, which leads to a safety goal violation. The use of the same signal for both decomposition functions is only permitted if the signal ASIL is same as the original ASIL of the safety goal.

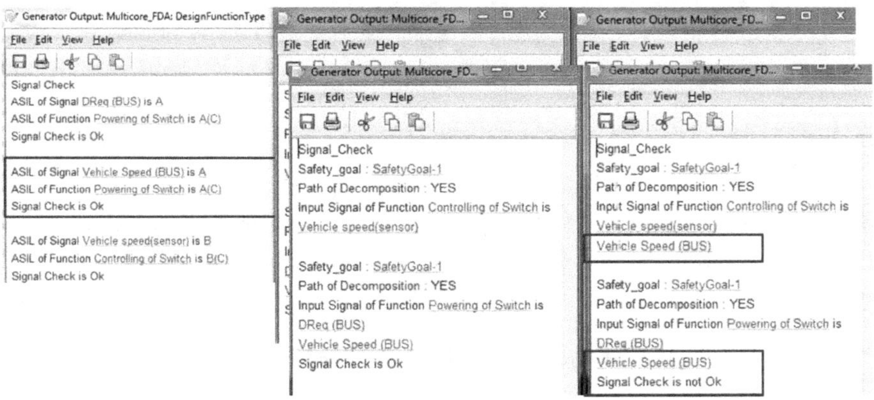

Figure 4.57: Use case: Signal check-nOk/Decomposition violation because of signal dependency

The script also checks for components of the identical type. The sensor signals that are used for the decomposed functions must be conditioned independently. The relationships between signals and peripheral elements are modeled using an allocation matrix, as shown in Fig. 4.58.

Afterwards, the corresponding hardware components for the signals are generated automatically based on the allocation matrix, as shown in Fig. 4.59. Finally, the shared resources of these decomposed signals are listed. It is not permitted to use the same hardware peripheral elements, e.g., the ADC, timer modules etc., for the conditioning of the signals which are used for the decomposed functions. There is only one exception which permits the use of the same elements for the decomposition, namely, the safeguarding of the hardware failure for these shared resources according to the original ASIL of the safety goal.

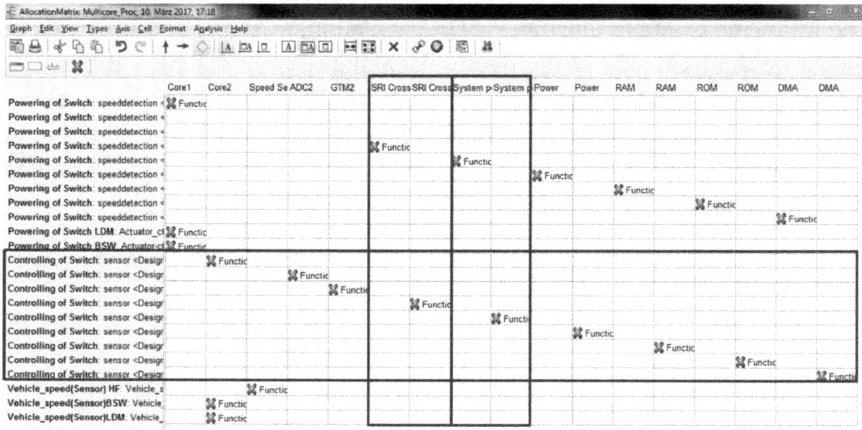

Figure 4.58: Use case: Signal check/Allocation between signals and hardware elements

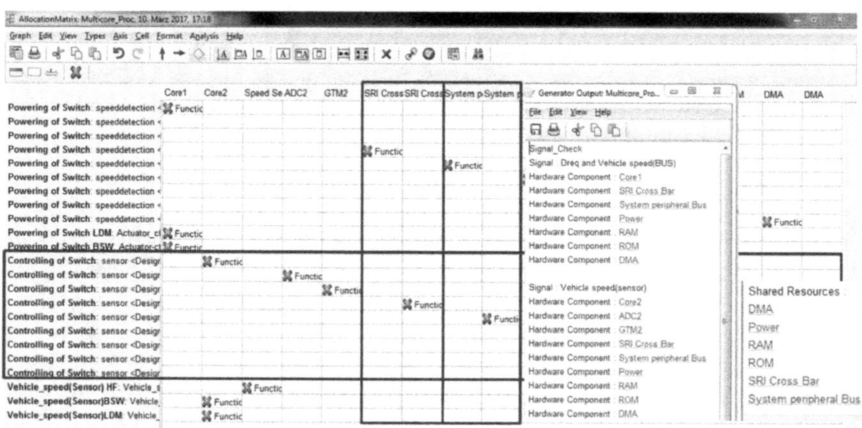

Figure 4.59: Use case: Signal check/Detection of shared resources

4.2.7 Conclusion

The presented approach and the developed automated model-based DFA in this research support the developers in creating the safety architecture and in providing evidence of sufficient independence between decomposed architectural elements.

The proposed model-based DFA resulting from this research enables the evidence of sufficient independence between decomposition paths and signals to be checked automatically. The automation of this step reduces the development effort because the decomposition path from the system architecture to the software and hardware architecture can be analyzed using this model-based approach instead of being analyzed manually. The traceability ensures that the signals can be checked in early design stage before they are used in a safety function. It is more efficient to be able to detect, in an earlier development phase, all of the signals that are not permitted to be used because of decomposition violations.

The other important advantage of this method is the traceability of the different steps. Once the user is familiar with the method, it is easy to gain an understanding of the overall architecture. Therefore, the approach enables the user to comprehend the system and safety architecture of an item very quickly. With the additional extensions of EAST-ADL regarding multicore processor-related elements, it is now possible to create the architectural design of multicore processors. The method also permits the modeling of a multicore processor with its hardware elements and software safety architecture, which is necessary to prove hardware and software independency. The extensions to EAST-ADL and the scripts developed for this method make sufficient transparency and traceability for the safety arguments possible, and support the entire safety process for multicore processors in a single solution. Within this approach, the EAST-ADL modeling, e.g., requirement modeling, dependency modeling and hardware modeling, is enhanced to be able to model the safety-critical functions with multicore processors. These extensions make it easy for the user to determine which requirements are implemented in which functions, and which functions are located on which core.

5 Conclusion and Outlook

This research outlines the various fail-operational safety architecture methods developed with consideration of domain ECUs containing multicore processors and describes the model-driven approaches for the development of the safety lifecycle and the automated DFA.

The methods presented in chapter 3 provide fail-operational system architecture and safety architecture for both conventional domains such as powertrains and for ADAS/AD systems in relation to the processing chain from sensors to actuators.

In the conventional domain context, this research investigates whether and how fault-tolerant systems such as the 2oo3 safety architecture can be realized with domain ECUs consisting of two multicore processors. The result of the research shows that the different ECUs can be combined in one domain ECU in a fail-operational manner. The main feature of this approach is the usage of a second multicore processor as a backup for the faulty processor or arithmetic core. The benefit of this method is the increase in system availability provided by the fail-operational architecture within the redundant cores.

In the ADAS/AD system context, this research provides various methods to improve the state of the art in the development of an ADAS/AD system. The methods in subchapter 3.4.1 show how a fail-operational system can be developed systematically, beginning with the relevant operating design domain. Some fault-tolerant methods already exist for other branches, such as rail and avionics. In comparison to these other branches, AD vehicle systems must be designed in a different way to handle the complexity of capturing the situational awareness with various sensors, processing large amounts of sensor data with high performance chips and controlling the vehicle in self-driving mode within fail-operational actuators. The currently available papers on fail-operational automotive systems do not deal in this context with the ADAS/AD specific fail-operational challenges.

B. Sari, *Fail-operational Safety Architecture for ADAS/AD Systems and a Model-driven Approach for Dependent Failure Analysis*, Wissenschaftliche Reihe Fahrzeugtechnik Universität Stuttgart, https://doi.org/10.1007/978-3-658-29422-9_5

This research concentrates on solving these challenges, especially with regard to the AD processing chain and domain ECUs consisting of high performance chips and multicore processors. The main advantage is the realization of the electronic systems of the vehicles, from sensor to actuator, in a fail-operational manner. The resulting fail-operational safety architecture methods depend on the safety and availability requirements of different SAE automation levels. The results of this research answer the question of which of the developed safety architecture methods is most suitable for a specific system in consideration of system requirements, operating design domain and automation level. The proposed methods for a fail-operational safety architecture provide a significant contribution to the development of ADAS/AD systems, especially for fully self-driving vehicles, so that these vehicles can be designed to be more safe, i.e., to achieve the minimum risk condition in every driving situation, to increase the availability and also to gain the acceptance of the customer for fully self-driving vehicles.

The alternative approach in subsection 3.5.3 provides different solutions for the application of the decomposition method for ADAS/AD systems concerning the relationship between different sensors and domain ECUs. These solutions show how an ASIL-D safety goal can be implemented in consideration of the ADAS/AD processing chain, which requires ASIL-D rated sensor information, ASIL-D rated perception algorithms, ASIL-D rated driving behavior functions and control of the actuators according to ASIL-D. The method resulting from this research in subsection 3.4.3 explains how the DFA from ISO 26262 can be extended for ADAS/AD systems while taking SOTIF-related technological insufficiencies into account. Additionally, the developed method in subsection 3.4.1.3 illustrates how a suitable fallback strategy is designed to achieve the minimum risk condition and to increase the availability in the case of a system failure or system limitations. In the development of this fallback strategy, two important aspects are considered, safety and availability. System safety is essential for AD systems, but availability also plays a significant role in fulfilling customer expectations. Therefore, these two facets are considered in the resulting generic intelligent fallback strategy.

In this research, the EAST-ADL standard, the EAST-ADL abstraction layer, the dependability model and the requirements model are extended in a way that enables efficient and consistent model-based development of the automotive system and safety architecture, and combines the system and safety development to achieve traceability and to show the relationship between the safety goals and the safety functions with consideration of ASIL. The presented model-driven approaches in chapter 4, for the development of the safety lifecycle and the developed automated model-based DFA, support the developers in creating the safety architecture and assist them in providing evidence of sufficient independence between decomposed architectural elements.

The model-based DFA proposed in this research enables an automatic check of the evidence of sufficient independence between the decomposition paths and the signals and also diverse paths of the ADAS/AD systems. The automation of this step reduces the time and effort spent on the development because the decomposition path from the system architecture to the software and hardware architecture can be analyzed using this model-based approach. The supported traceability ensures that the usage of the signals can be checked in an early design stage before the signals are used in a safety function. The capability, at an early stage of the development, of finding all the signals which are not permitted to be used because of a decomposition violation, increases the efficiency of the process. The other important advantage of this method is the traceability of the various steps. Once the user is familiar with the method, it is easy to gain an understanding of the overall architecture, so the approach enables the user to grasp the system and safety architecture of a specific item quickly. With the additional extensions to EAST-ADL regarding multicore processor-related elements, it is now possible to create the architectural design for multicore processors. The enhancements to EAST-ADL also permit the modeling of a multicore processor with the hardware elements and the modeling of the software safety architecture, which are necessary to prove hardware and software independency. Finally, the extensions and developed scripts make it possible to achieve sufficient transparency and traceability for the safety arguments and to support the whole safety process for multicore processors in a single solution, throughout the development of the system, hardware and software architecture.

The purpose of this research, to advance the fail-operational safety architecture methods regarding the ADAS/AD processing chain and model-based DFA method for self-driving vehicle systems, has been served by the results. In further research endeavors, the fail operational aspects of software, for example diverse implementation of algorithms, can be extended with respect to the independence requirements of the decomposition approach and also the main/redundant path. There are still many challenges concerning the development of autonomous vehicles to be solved. Some of these challenges are; How can the basic software in domain ECUs be designed to fulfill the high level safety requirements? How can machine learning algorithms be trained diversely? How can the variables of these algorithms be kept updated after the detection of new objects? What are the possible validation methods for machine learning algorithms? How can a suitable validation target for self-driving vehicles be calculated for the purpose of decreasing the unknown unsafe scenarios to the level which is acceptable for the release of the functions? Additionally, in the future, the model-based DFA approach can continue to be extended in such a way that the safety mechanisms of the detected common causes are considered by performing the DFA. This would enable an automatic check of whether the identified common causes lead to a safe state, and continue the work toward the ultimate goal of driving safety.

Bibliography

[1] Autoindustrie treibt Chipnachfrage an. http://www.pwc.de/de/automobilindustrie/autoindustrie-treibt-chipnachfrage-an.html. accessed: 30.01.2017.

[2] EAST-ADL Overview. http://www.maenad.eu/public/conceptpresentations/EAST-ADL_WhitePaper_M2.1.12.pdf. accessed: 10.12.2018.

[3] EAST-ADL Overview and Structure. http://www.atesst.org/home/liblocal/docs/ConceptPresentations/01_EAST-ADL_OverviewandStructure.pdf. accessed: 10.12.2018.

[4] EAST-ADL Specification. http://www.east-adl.info/Specification.html. accessed: 10.12.2018.

[5] HiP-HOPS, Automated Fault Tree, FMEA and Optimisation Tool. http://www.hiphops.eu/index.php/the-manual. accessed: 10.12.2018.

[6] *IEC61508: Functional safety of electrical/electronic/programmable electronic safety-related systems*. International Electrotechnical Commission, 2005.

[7] EAST-ADL Analysis Level. http://www.atesst.org/home/liblocal/docs/ConceptPresentations/03_EAST-ADL_Analysis_Level.pdf, 2010. accessed: 10.12.2018.

[8] EAST-ADL Design Level. http://www.atesst.org/home/liblocal/docs/ConceptPresentations/04_EAST-ADL_Design_Level.pdf, 2010. accessed: 10.12.2018.

[9] EAST-ADL Implementation Level. http://www.atesst.org/home/liblocal/docs/ConceptPresentations/05_EAST-ADL_Implementation_Level.pdf, 2010. accessed: 10.12.2018.

[10] EAST-ADL Vehicle Level. http://www.atesst.org/home/liblocal/docs/ConceptPresentations/02_EAST-ADL_Vehicle_Level.pdf, 2010. accessed: 10.12.2018.

B. Sari, *Fail-operational Safety Architecture for ADAS/AD Systems and a Model-driven Approach for Dependent Failure Analysis*, Wissenschaftliche Reihe Fahrzeugtechnik Universität Stuttgart, https://doi.org/10.1007/978-3-658-29422-9

[11] *Critical Reasons for Crashes Investigated in the National Motor Vehicle Crash Causation Survey*. National Highway Traffic Safety Administration, February 2015.

[12] *Global Status Report on Road Safety 2015*. World Health Organization, 2015.

[13] *SAE J3016. Taxonomy and Definitions for Terms Related to Driving Automation Systems for On-Road Motor Vehicles*. SAE International, September 2016.

[14] *ISO FDIS 26262 2nd. Edition: Road Vehicles – Functional Safety*. The International Organization for Standardization (ISO), 2nd Edition, 2018.

[15] *ISO/PAS 21448: Safety of the Intended Functionality*. The International Organization for Standardization (ISO), 2018.

[16] AUTOSAR. AUTOSAR (AUTomotive Open System ARchitecture). http://www.autosar.org/. accessed: 10.12.2018.

[17] F. K. Bapp. *Adaptives Monitoring für Mehrkernprozessoren in eingebetteten sicherheitskritischen Systemen*. PhD thesis, Karlsruhe Institute of Technology, 2017.

[18] H. Blair-Smith. *Space shuttle fault tolerance: Analog and digital teamwork*. 2009.

[19] H. Blom, H. Lönn, F. Hagl, and Y. Papadopoulos, editors. *EAST-ADL – An Architecture Description Language for Automotive Software-Intensive Systems – White Paper Version 2.1.12*, January 2013.

[20] R. N. Charette. This car runs on code. http://www.real-programmer.com/interesting_things/IEEE%20SpectrumThisCarRunsOnCode.pdf. accessed: 10.12.2018.

[21] P. Cuenot, D. Chen, S. Gerard, and H. L. et al., editors. *Managing Complexity of Automotive Electronics Using the EAST-ADL*, 12th IEEE International Conference on Engineering Complex Computer Systems, 2007.

[22] E. Dubrova. *Fault-Tolerant Design: An Introduction*. 2008.

[23] Ford. A Matter of Trust: Ford Safety Assessment Report For Self-Driving Vehicle Development. https://media.ford.com/content/fordmedia/fna/us/en/news/2018/08/16/a-matter-of-trust-ford-releases-safety-assessment-report.html, 2018. accessed: 18.08.2018.

[24] GM. Self-Driving Safety Report. https://www.gm.com/mol/ selfdriving.html, 2018. accessed: 18.08.2018.

[25] A. Hachiga, K. Akita, and Y. Hasegawa. The design concepts and operational results of fault-tolerant computer systems for the shinkansen train control. In *Twenty-Fifth International Symposium on Fault-Tolerant Computing, 1995, ' Highlights from Twenty-Five Years'.*, pages 383–, June 1995.

[26] A. Kohn, M. Käßmeyer, R. Schneider, A. Roger, C. Stellwag, and A. Herkersdorf, editors. *Fail-operational in safety-related automotive multi-core systems*. Industrial Embedded Systems (SIES), 10th IEEE International Symposium, 2015.

[27] A. Kohn, R. Schneider, A. Vilela, and A. Roger, editors. *Architectural Concepts for Fail-Operational Automotive Systems*. SAE Technical Paper, apr 2016.

[28] P. Koopman and M. Wagner, editors. *Autonomous Vehicle Safety: An Interdisciplinary Challenge*, Pittsburgh, U.S.A, june 2016. IEEE Intelligent Transportation Systems Magazine.

[29] P. Koopman and M. Wagner, editors. *Toward a Framework for Highly Automated Vehicle Safety Validation*, Detroit, U.S.A, apr 2018. SAE International, WCX™ 18: SAE World Congress Experience.

[30] M. Maurer, J. C. Gerdes, B. Lenz, and H. Winner. *Autonomous Driving*. 2015.

[31] G. Moore, editor. *Understanding Moore's Law: Four Decades of Innovation*, Chemical Heritage Foundation. pp. 67–84. ISBN 0-941901-41-6, San Jose, CA, U.S.A., March 2018.

[32] NASA. *NASA Software Safety Guidebook*. 2004.

[33] NHTSA. Federal Automated Vehicles Policy. Accelerating the Next Revolution in Roadway Safety. https://www.transportation.gov/sites/dot.gov/files/docs/AV%20policy%20guidance%20PDF.pdf, 2016. accessed: 18.08.2018.

[34] OECD and ITF. *Automated and Autonomous Driving: Regulation under uncertainty*. 2015.

[35] SAFE. Safe Automotive Software Architecture (SAFE). http://www.safe-project.eu/. accessed: 10.12.2018.

[36] B. Sari, editor. *Model-Based Dependent Failure Analysis*, 9th International Annual Conference ISO 26262, Stuttgart, Germany, October 2017.

[37] B. Sari and A. Das, editors. *Extending SOTIF Application for higher Automated Driving*, SAE International: WCX 18: SAE World Congress Experience, Detroit, U.S.A., April 2018.

[38] B. Sari and A. Das, editors. *SOTIF for higher Automated Driving*, safe.tech conference 2018, München, Germany, April 2018.

[39] B. Sari, S. Gohl, and R. Geiger, editors. *Ein modellgetriebener Ansatz zur Beherrschung von sicherheitskritischen Funktionen im E-Antrieb*, safe.tech conference 2016, München, Germany, April 2016.

[40] B. Sari and H.-C. Reuss, editors. *A model-driven approach for the development of safety-critical functions using modified Architecture Description Language (ADL)*, Electrical Systems for Aircraft, Railway, Ship propulsion and Road Vehicles and International Transportation Electrification Conference (ESARS-ITEC), Toulouse, Frankreich, November 2016.

[41] B. Sari and H.-C. Reuss, editors. *Fail-operational Safety Architecture for Domain-ECUs with Multicore Processors*, The 30th International Electric Vehicle Symposium and Exhibition, Stuttgart, Germany, October 2017.

[42] B. Sari and H.-C. Reuss. Model-based Development of Safety-critical Functions and ISO 26262 Work Products using modified EAST-ADL.

Advances in Science, Technology and Engineering Systems Journal, 2(3):1252–1259, 2017.

[43] B. Sari and H.-C. Reuss, editors. *A Model-Driven Approach for Dependent Failure Analysis in Consideration of Multicore Processors Using Modified EAST-ADL*, Detroit, U.S.A, mar 2017. SAE International, WCX™ 17: SAE World Congress Experience.

[44] B. Sari and H.-C. Reuss, editors. *Fail-Operational Safety Architecture for ADAS Systems Considering Domain ECUs*, Detroit, U.S.A, apr 2018. SAE International, WCX™ 18: SAE World Congress Experience.

[45] A. Schnellbach. *Fail-operational automotive systems*. PhD thesis, Graz University of Technology, 2016.

[46] Synligare. Synligare (literally "more visible" in swedish). http:// synligare.eu/HomePage.html/. accessed: 10.12.2018.

[47] D. Takahashi, editor. *A decade later, he revised what had become known as Moore's Law: The number of transistors on a chip would double every two years*, Seattle Times, San Jose, CA, U.S.A., April 2015.

[48] WAYMO. Waymo safety report. https://waymo.com/safety/, 2017. accessed: 18.08.2018.

[49] H. Winner, S. Hakuli, F. Lotz, and C. Singer. *Handbuch Fahrerassistenzsysteme*. 2015.

[50] D. Wyatt and M. Tooley. *Aircraft Electrical and Electronic Systems: Principles, Maintenance and Operation*. 2009.

Printed in the USA
CPSIA information can be obtained
at www.ICGtesting.com
CBHW062203120824
13105CB00003B/96